D1785860

A CLINICAL GUIDE TO

ANAEROBIC
INFECTIONS

Also Available from Star Publishing Company:

WADSWORTH ANAEROBIC BACTERIOLOGY MANUAL
PRINCIPLES AND PRACTICE OF ANAEROBIC BACTERIOLOGY

A CLINICAL GUIDE TO
ANAEROBIC INFECTIONS

Sydney M. Finegold, M.D.
Ellen Jo Baron, Ph.D.
Hannah M. Wexler, Ph.D.

Wadsworth Anaerobe Laboratory
Los Angeles, California

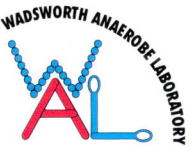

Hans J.R. Tester, M.D.
Medical Editor

PUBLISHING COMPANY

Published by Star Publishing Company
Publisher and Managing Editor: Stuart A. Hoffman
Medical Editor: Hans J.R. Tester, M.D.
Drawings by Jody Fulks-Sjogren and Princeton Graphic Productions, Inc.
Cover by Douglas B. Hurd and Princeton Graphic Productions, Inc.

The authors and publisher have exerted efforts to ensure that procedures, reference ranges, and drug selection and dosages set forth in this text are in accord with current recommendations and practice at the time of publication. However, in view of ongoing microbiology and medical research, changes in government regulations, and the constant flow of information relating to microbiology, drug therapy, and drug reactions, the reader must check the latest available information for each drug for any change in indications and dosage and for added warnings and precautions. This is particularly important when the recommended agent is a new or infrequently employed drug.

STAR Publishing Company
P.O. Box 68
Belmont, CA 94002
United States of America

ISBN: 0-89863-162-9

94 93 92 91   10 9 8 7 6 5 4 3 2 1
Printed in Singapore

Table of Contents

Preface

Anaerobic and mixed aerobic-anaerobic infections are quite common and often serious. Appropriate treatment of these infections depends upon an understanding of the pathogens responsible. Reliable bacteriology, unfortunately, often is not obtained on patients with such infections, for a variety of reasons. Unless specimens are collected and transported optimally, results of cultures may be unreliable or even misleading. Some clinical microbiology laboratories are not well set up to carry out good anaerobic work. Even when everything else is appropriate, cost considerations now prevent most laboratories from doing optimal cultures for anaerobes. Many times, to be sure, there is no real advantage or necessity to do careful, detailed anaerobic bacteriology, particularly in nonteaching hospitals, when the usual bacterial flora of certain types of infections is well known.

Another dilemma is that it may take extended periods of time to isolate the various components of a complex, mixed infection in pure culture and to identify them. Clinicians need to make prompt decisions regarding therapy for their patients. Accordingly, therapy necessarily is empirical in such situations. In order to provide rational and informed empiric therapy, the clinician must determine the exact *nature* of the infection to be treated and must know the *typical bacteriology* of such an infection and the *usual antimicrobial susceptibility patterns* (in his/her hospital) for the organisms involved. This type of information, of course, should be supplemented by Gram stains of appropriate specimens, by aerobic cultures (which can generally give reliable information within 24 hours), and by knowledge of how the infecting flora may have been modified by pathophysiologic events and prior antimicrobial therapy. Perforated peptic ulcer, for example, usually involves a simple, sparse flora derived from swallowed oropharyngeal organisms. In the presence of bleeding or obstruction, however, the gastric and duodenal flora may be quite profuse and diverse, even resembling colonic flora. Antimicrobial agents eliminate susceptible organisms, but may select out resistant strains; therefore, knowledge of the spectrum of activity of various antimicrobial drugs is important.

The purpose of this guide is to present, in succinct and graphic fashion, information about the principal anaerobic or mixed infections (their etiology, underlying problems, clinical features, diagnosis, and therapy) along with information on anaerobes as normal flora, the incidence of anaerobes in infections of various types, sampling and transport methodology, representative Gram stains from clinical specimens, susceptibility test methods and data for anaerobes, and a guide to therapy.

It is our hope that the material presented here will be useful to clinicians in providing their patients with optimal empiric therapy.

Acknowledgments

The authors are grateful to Jeffrey Tuttle, of Symedco, for the origination of the concept for this publication and Anna Holland for diligently and patiently undertaking the word processing tasks in preparation of the manuscript. We also thank the following friends and colleagues for their valuable photographic contributions.

Figures 1, 37, 44, 47, 66	Samuel E. Wilson, M.D.
Figures 4, 16, 20, 26, 29, 48, 49, 58, 62, 80	*Scope Monograph on Anaerobic Infections*, by Sydney M. Finegold, M.D. and Vera L. Sutter, Ph.D. The Upjohn Co. Reprinted with permission.
Figure 5	Russell Klein, M.D.
Figures 9, 10, 52, 53, 55, 70, 86	A. Trevor Willis, M.D.
Figures 11, 15, 17, 89	John G. Bartlett, M.D.
Figures 18, 61	Donald Poretz, M.D.
Figures 19, 96, 97	Peter M. Rose
Figures 24, 25	Sydney M. Finegold, M.D. (Reprinted by permission of *Hospital Practice* 24:103, 1989)
Figure 51	Larry C. Ford, M.D.
Figure 54	Ellie J.C. Goldstein, M.D.
Figures 56, 59	Ian Phillips, M.D., and Susannah J. Eykyn, M.D.
Figures 64, 65	D.S. deJongh, M.D., J.P. Smith, M.D., and G. W. Thomas, M.D. (*JAMA* 200:557-559, Copyright 1967, American Medical Association)
Figure 67	Haragopal Thadepalli, M.D.
Figures 68, 69, 80	Martin C. McHenry, M.D.
Figures 71, 81	Reproduced by permission from Baron, Ellen Jo, and Finegold, Sydney M.: Bailey & Scott's *Diagnostic Microbiology*, ed 8, St. Louis, 1990, The C.V. Mosby Co.

Figures 72, 73   C.R. Baxter, M.D. (Surgical Management of Soft Tissue Infections. *The Surgical Clinics of North America* 1972, Vol. 52, p. 1483)   Reprinted by permission

Figures 82, 83   Charles V. Sanders, M.D.

Figure 87   Stephen S. Arnon, M.D.

Figure 88   The Upjohn Company

Figure 90   King K. Holmes, M.D., Ph.D.

SECTION 1

INTRODUCTION TO ANAEROBIC INFECTION

SECTION 1

INTRODUCTION TO ANAEROBIC INFECTION

Pathogenesis of Anaerobic Infection

Anaerobic infections may arise in different ways and have a variety of clinical presentations with abscess formation and tissue necrosis especially common. These infections are derived from the normal flora of the oronasopharynx, skin, bowel or female genital tract. Determinants of mixed aerobic/anaerobic infection include the inoculum size, virulence factors produced by the organisms present, host defense mechanisms and factors predisposing to infection.

Virulence Factors in Anaerobes

The three major virulence factors in anaerobes are the ability to adhere to or invade epithelial surfaces, the production of toxins or enzymes that play a pathogenic role, and surface constituents of the organisms such as capsular polysaccharide or lipopolysaccharide.

The ability to adhere to epithelial cells is vital to establishment of colonization or infection. Both *Bacteroides melaninogenicus* and *Fusobacterium nucleatum* are known to adhere to crevicular epithelium in the oral cavity, with the former organism showing an ability to attach to certain gram-positive organisms in vitro. *Porphyromonas gingivalis*, thought to be an important organism in human periodontal disease, possesses fimbriae that facilitate attachment. The three different types of structures shown to be responsible for the adherence of *Bacteroides fragilis* to various epithelial structures are the capsule, negative-staining structures consistent with pili, and lectin-like adhesins. Binding and degrading of human fibrinogen by *P. gingivalis* may mediate colonization with this organism in the gingival crevice.

Numerous toxins and enzymes play a role in bacterial virulence. Enzymes believed to be important with regard to invasion include phospholipase A, collagenase and hyaluronidase. Superoxide dismutase permits anaerobic bacteria to survive exposure to oxygen. *Clostridium perfringens* serves as a model for toxin production among anaerobes. Its major toxin is alpha toxin, a phospholipase C. This enzyme hydrolyzes lecithin and sphingomyelin in cell membranes of a number of cell types including red blood cells, platelets, endothelial cells and muscle cells. This toxin, and two others produced by *C. perfringens*, affects capillary permeability. This organism also produces a collagenase. Other toxins and enzymes produced by anaerobes include neuraminidase, deoxyribonuclease, phosphatase, heparinase, leukocidin, hemolysins, hemagglutinins, lysophospholipase, proteinases, protease, sulfatase, sialidase, various enterotoxins, tetanus neurotoxin, tetanolysin, and botulinal toxin.

Surface constituents include capsules and lipopolysaccharide or endotoxin. The capsular polysaccharide of *B. fragilis*, free of other components of the bacterial cell, is capable of inducing abscess formation.

Host Defense Mechanisms

Certain of the gram-negative anaerobic bacilli are killed directly by serum complement. Random migration of polymorphonuclear leukocytes does not differ significantly under aerobic and anaerobic conditions; however, anaerobes attract polymorphonuclear leukocytes into their immediate area by activation of complement and by direct mechanisms. In these cells both oxidative and nonoxidative mechanisms contribute to killing of the anaerobes. It is likely, also, that anaerobes are susceptible to killing by macrophages.

Acquired immunity involves both humoral and cell-mediated immune mechanisms. Circulating antibody and complement protect against bacteremia associated with experimental intra-abdominal infection, and T lymphocytes contribute to resistance against abscess formation.

Anaerobes may exert adverse effects on humoral and cellular host defense mechanisms. Some anaerobes may compete with nonanaerobic bacteria for serum opsinins and, under certain conditions in vitro, may directly depress the function of polymorphonuclear leukocytes, macrophages and lymphocytes.

Factors Predisposing to Infection

Anaerobes are prevalent as indigenous flora on all mucosal surfaces. The mucosal barrier may be disrupted by surgery, trauma or various disease states, thus affording an opportunity for these organisms to penetrate deeper tissues and to set up infection. In other cases (e.g., aspiration pneumonia), anaerobic bacteria from a site of normal carriage (oropharynx in this case) may move into another area normally free of organisms to produce infection at that site. Tissue necrosis or poor blood supply lowers the oxidation-reduction potential, favoring the growth of anaerobic organisms. Vascular disease, cold, shock, trauma, surgery, foreign bodies, malignancy, edema and gas production by bacteria, therefore, may significantly predispose to anaerobic infection, as may previous infection with nonanaerobic bacteria. Antimicrobial agents to which anaerobes are notably resistant (e.g., aminoglycosides) may facilitate anaerobic infection. Anaerobic organisms that are more aerotolerant (e.g., those producing superoxide dismutase) are more likely to survive after the normally protective mucosal barrier is broken and until conditions are satisfactory for multiplication and invasion by these organisms. As anaerobes multiply, they maintain their own reduced environment by means of their end-products of fermentative metabolism.

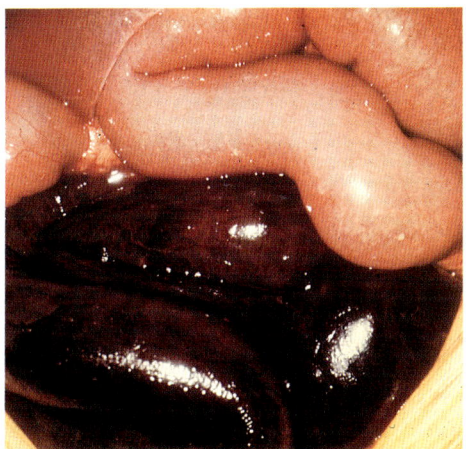

Figure 1.
Strangulation obstruction of bowel.
Poor blood supply, tissue damage, and
intestinal flora combine to favor anaero-
bic or mixed infection (peritonitis and/or
intra-abdominal abscess).

Anaerobes as Normal Flora

A knowledge of the anaerobes present as normal flora at various sites in the body is important since virtually all anaerobes involved in infection originate from this flora. When infections arise at a certain site, knowing which organisms predominate there assists the clinician in choosing appropriate drugs for initial therapy. The nature of an organism involved in bacteremia of uncertain source may point to the likely portal of entry. The following table lists the types of anaerobes found at different body sites.

Table 1.   ANAEROBES AS NORMAL FLORA

Location	Gram-Positive Spore-Forming Clostridium	Gram-Positive Non-Spore-Forming			Gram-Negative Non-Spore Forming			
		Actinomyces	Propionibacterium	Cocci	Bacteroides	Fusobacterium	Bilophila	Cocci
Skin	0	0	2	1	0	0		0
Upper respiratory tract[a]	0	1	1	1	1	1		1
Mouth	±	1	±	2	2	2		2
Intestine	2	±	±	2	2	±	1	±
External genitalia	0	0	U	1	1	1		0
Urethra	±	0	0	±	1	1		U
Vagina	±	0	1	2	1	±		1

U, unknown; 0, not found or rare; ±, Irregular; 1, usually present; 2, usually present in large numbers.
[a]Includes nasal passages, nasopharynx, oropharynx, and tonsils.

Modified from Sutter VL, Citron DM, Edelstein MAC, Finegold SM, *Wadsworth Anaerobic Bacteriology Manual*, 4th ed. Star Publ. Co., Belmont, CA, 1985

Predominant anaerobes of the normal flora

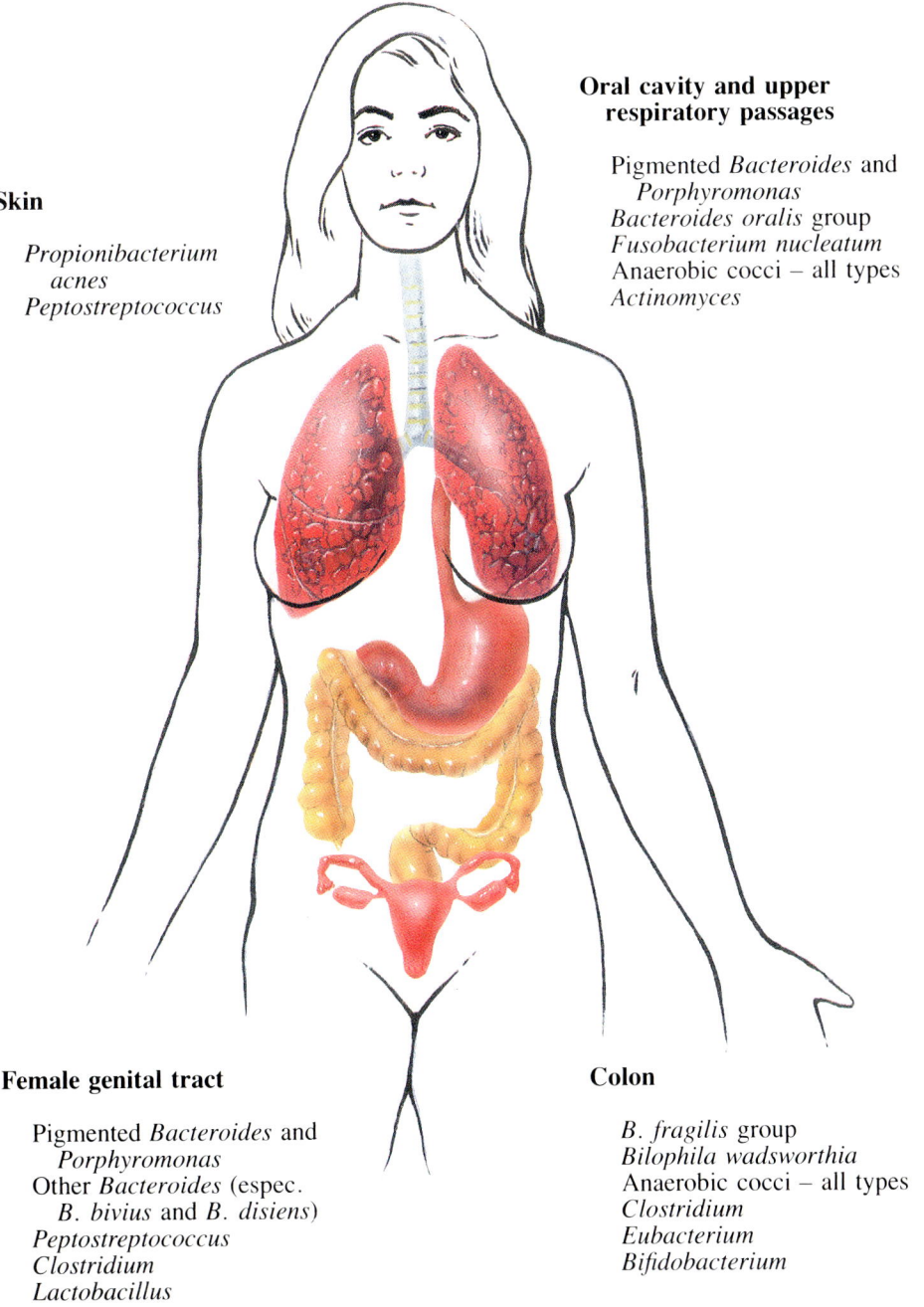

Oral cavity and upper respiratory passages

Pigmented *Bacteroides* and
   Porphyromonas
Bacteroides oralis group
Fusobacterium nucleatum
Anaerobic cocci – all types
Actinomyces

Skin

*Propionibacterium
   acnes
Peptostreptococcus*

Female genital tract

Pigmented *Bacteroides* and
   Porphyromonas
Other *Bacteroides* (espec.
   B. bivius and *B. disiens*)
*Peptostreptococcus
Clostridium
Lactobacillus*

Colon

B. fragilis group
Bilophila wadsworthia
Anaerobic cocci – all types
*Clostridium
Eubacterium
Bifidobacterium*

Figure 2.

7

Common Anaerobic Pathogens

Following is a list of the anaerobic bacteria that are important pathogens, either because they are commonly encountered in clinical infection, because they produce serious infection, or both.

Gram-Negative Bacilli	Gram-Positive Spore-Forming Bacilli
Bacteroides fragilis group[a] (especially *B. fragilis* and *B. thetaiotaomicron*)	*Clostridium perfringens*
	C. ramosum
Pigmented *Bacteroides* and *Porphyromonas*[b]	*C. septicum*
B. oris, B. buccae (B. ruminicola)	*C. novyi*
B. oralis group *(B. oralis, B. veroralis,*	*C. histolyticum*
B. buccalis, B. oulorum)	*C. sporogenes*
B. ureolyticus group *(B. ureolyticus, B. gracilis,*	*C. sordellii*
Wolinella spp., *Campylobacter concisus)*	*C. bifermentans*
B. disiens	*C. fallax*
Fusobacterium nucleatum	*C. difficile*
F. necrophorum	*C. innocuum*
F. gonidiaformans	*C. botulinum*
F. naviforme	*C. tetani*
F. mortiferum	
F. varium	
Bilophila wadsworthia	

Gram-Positive Cocci	Gram-Positive Non-Spore-Forming Bacilli
Peptostreptococcus (especially *P. magnus,* *P. asaccharolyticus, P. prevotii, P. anaerobius,* *P. intermedius,*[c] *P. micros)*	*Actinomyces (A. israelii, A.* *meyeri,* *A. naeslundii,* *A. odontolyticus,* *A. viscosus)*
Microaerophilic streptococci[c]	*Propionibacterium propionicus (Arachnia propionica)*
	Propionibacterium acnes
	Bifidobacterium dentium (eriksonii)

[a] Includes *B. fragilis, B. thetaiotaomicron, B. distasonis, B. vulgatus, B. ovatus, B. uniformis, B. caccae* (B 3452A) and others.
[b] Includes *B. melaninogenicus, P. asaccharolyticus, P. gingivalis, B. intermedius, B. corporis, B. denticola, B. loescheii, P. endodontalis, B. bivius* and others.
[c] Not true anaerobes

Prevalence of Anaerobes in Infection

Certain specific anaerobes predominate in the majority of infections in which these organisms are involved. The *B. fragilis* group, the pigmented anaerobic gram-negative bacilli, *F. nucleatum*, *C. perfringens* and *C. ramosum*, and peptostreptococci, taken together as a group, account for half to two-thirds of clinically significant anaerobic bacterial isolates.

The table on pages 10-11 provides details on the frequency of isolation of various anaerobes from specific specimen sources or infection types in our experience at the Wadsworth Veterans Administration Medical Center.

Infections Commonly Involving Anaerobes

While anaerobes may cause any type of infection in humans, there are certain infections in which anaerobic bacteria are more commonly or characteristically found. The table on pages 12-13 lists the incidence of anaerobes in these infections, and the percent of these infections that yield only anaerobes.

Table 2. PREVALENCE OF ANAEROBES IN INFECTIONS

Anaerobe	Blood	CNS[b]	Head and Neck Infections	Dental	Bites Human	Bites Animal	TTA[c] and Pleural Fluid	Miscellaneous Soft Tissue Infections Above Waist	Below Waist	Intra-Abdominal Infections	Perirectal Abscess	Ulcers Decubitus	Foot	Osteomyelitis
Number of positive specimens surveyed	175	19	75	9	25	33	196	79	135	185	34	54	222	49
Total isolates	226	32	349	35	72	81	656	203	375	763	158	212	645	139
Average number of anaerobes per specimen	1.3	1.6	4.6	3.9	2.9	2.5	3.3	2.6	2.7	4.1	4.6	3.9	2.9	2.8
B. fragilis	33	0	0	0	0	3	4	5	35	54	68	50	30	12
B. thetaiotaomicron	9	0	1	0	0	0	1	0	15	30	32	26	6	6
Other bile-resistant Bacteroides[d]	13	0	3	0	0	0	2	0	14	50	62	48	12	4
B. melaninogenicus-loescheii-denticola	1	0	40	11	20	6	23	4	4	5	6	4	5	6
B. intermedius-corporis group	0	5	49	11	48	6	30	18	5	12	6	6	8	2
Porphyromonas asaccharolytica	1	5	5	0	0	6	3	10	9	6	26	17	10	2
Other pigmented Bacteroides or Porphyromonas spp.	0	0	12	0	0	3	6	3	5	12	9	9	7	10
B. ureolyticus group	1	5	7	0	4	0	9	9	6	5	12	2	5	8
Other bile-sensitive Bacteroides spp.[e]	3	5	72	22	48	12	47	29	9	15	29	11	14	12
Other Bacteroides[f]	5	11	41	11	20	24	33	42	14	24	21	35	20	24

F. nucleatum	2	32	40	0	28	15	29	22	5	13	9	2	4	6
F. necrophorum	7	5	1	22	0	0	3	4	0	4	0	0	1	2
F. mortiferum-varium group	1	0	0	0	0	0	0	0	1	4	3	2	1	0
Other Fusobacterium[g]	1	0	5	11	0	39	10	4	2	5	3	2	3	0
Other gram-negative bacilli	1	0	1	0	0	3	3	1	1	0	3	0	1	0
Peptostreptococcus spp.[h]	7	21	85	178	48	21	33	62	98	51	68	107	111	133
Veillonella	1	5	33	33	40	9	26	11	7	7	12	7	5	2
Acidaminococcus, Megasphaera, unidentified GNC	1	0	4	0	0	0	1	0	1	2	6	0	0	0
Clostridium perfringens	5	0	0	0	0	3	5	4	5	15	6	6	2	0
Other Clostridium spp.	21	21	0	0	4	0	5	1	6	44	59	39	7	2
Actinomyces spp.	0	32	8	44	12	21	14	6	9	3	0	9	6	10
Bifidobacterium spp.	1	0	3	0	0	0	5	0	1	2	0	0	0	4
Lactobacillus spp.	1	0	17	22	4	0	15	6	3	12	3	6	5	2
P. acnes	3	37	7	0	4	33	4	4	4	6	3	4	5	16
Other Propionibacterium spp.	0	0	3	11	4	21	2	5	2	1	0	0	2	0
Eubacterium spp.	7	0	25	11	4	18	23	11	16	37	18	13	12	16

[a] Numbers in this table represent numbers of isolates per 100 specimens containing anaerobes. Incidence of Specific Anaerobes in Various Infections: Wadsworth VA Medical Center, 1973-1983.

[b] CNS, Central nervous system.

[c] TTA, Transtracheal aspirates.

[d] B. distasonis, B. vulgatus, B. ovatus, B. uniformis, B. splanchnicus, B. eggerthii, and "B. fragilis group—no good fit."

[e] B. oris, B. buccae, B. oralis, B. disiens, B. buccalis, B. verordis.

[f] Includes nonspeciable Bacteroides.

[g] Includes nonspeciable Fusobacterium.

[h] Includes strains formerly identified as Peptococcus.

From Sutter VL, Citron DM, Edelstein MAC, Finegold SM. Wadsworth Anaerobic Bacteriology Manual. 4th ed., Star Publ. Co., Belmont, CA. 1985.

Table 3. INFECTIONS COMMONLY INVOLVING ANAEROBES

Infection	Incidence (%)	Anaerobes only (%)[a]
Bacteremia	20	80
Bacteremia secondary to tooth extraction	84	84
Ocular infections	38	23
Corneal ulcers	7	82
Central nervous system		
brain abscess	89	50-70
extradural or subdural empyema	10	[b]
Head and neck		
chronic sinusitis	52	82
acute sinusitis	7	
chronic otitis media	33-59	10
Cholesteatoma	92	9
Neck space infections	100	75
Wound infection following head and neck surgery	95	0
Peritonsillar abscess	76	21
Bite wounds	47	3
Dental and oral		
Orofacial, of dental origin	94	40
Root canal infection	95-100	30-70
Periodontal abscess	100	0
Dental abscess, endodontic origin	90-100	65-75
Thoracic		
aspiration pneumonia[c]	62-100	30-50[d]
lung abscess	85-93	50-75
bronchiectasis	76	33
empyema (nonsurgical)	62	50
Abdominal		
intra-abdominal (general)	81-94	10-35
appendicitis with peritonitis	96	1
liver abscess	52	33

Infection	Incidence (%)	Anaerobes only (%)[a]
Abdominal (cont.)		
other intra-abdominal infection (postsurgery)	93	17
wound infection following bowel surgery	e	b
biliary tract	41-45	0-2
Obstetric-gynecologic		
miscellaneous types	72-100	33
pelvic abscess	88	50
vulvovaginal abscess	75	25
vaginal cuff abscess	98	3
septic abortion, sepsis	63-67	b
pelvic inflammatory disease	25-48	7-14
Soft tissue and miscellaneous		
nonclostridial crepitant cellulitis	75	8
pilonidal sinus	73+	b
diabetic foot ulcers	95	5
infected diabetic gangrene (deep tissue culture)	85	9
soft tissue abscesses	60	25
cutaneous abscesses	62	20
decubitus ulcers with bacteremia	63	b
osteomyelitis	40	10
gas gangrene (clostridial myonecrosis)	100	b
breast abscess (subareolar)	e	b
perirectal abscess	e	b

---

[a] Percent cultures positive for anaerobes yielding only anaerobes.

[b] Blank area under proportion of cultures positive for anaerobes only indicates that no data are available.

[c] Aspiration pneumonia occurring in the community rather than in the hospital involves anaerobes to the exclusion of aerobic or facultative forms two-thirds of the time.

[d] 82% of 28 cultures yielding heavy growth of one or more organisms had only anaerobes present.

[e] Majority of cases yield anaerobes.

Modified from Sutter VL, Citron DM, Edelstein MAC, Finegold SM. *Wadsworth Anaerobic Bacteriology Manual*, 4th ed., Star Publ. Co., Belmont, CA, 1985.

SECTION 2

CLINICAL INFECTIONS

Table of Contents for Clinical Infections

SECTION 2

CLINICAL INFECTIONS

This section will review the major infections involving anaerobes. Some clinical clues to the presence of an anaerobic infection are given in the table below. These clues are not specific, except for the odor, but when two or more are present, anaerobes should be considered potential pathogens.

Table 4. CLINICAL CLUES TO ANAEROBIC INFECTIONS

1. *Foul or putrid odor to specimen*

2. *Location of infection in proximity to a mucosal surface*

3. *Infections secondary to human or animal bite*

4. *Gas in specimen*

5. *Tissue necrosis*

6. *Previous therapy with aminoglycoside antibiotics (such as gentamicin or amikacin) in the absence of concomitant effective coverage vs. anaerobes*

7. *Black discoloration of blood-containing exudates; these exudates may fluoresce red under ultraviolet light (infections involving pigmented* Bacteroides *or* Porphyromonas)

8. *Presence of "sulfur granules" in discharges (actinomycosis)*

9. *Unique morphology on Gram stain*

10. *Failure of organisms seen on Gram stain of original exudate to grow aerobically*

This section is intended to aid the clinician in the diagnosis and management of anaerobic infections. The infections are, for the most part, listed anatomically from head to toe. When appropriate, each infection includes sections on etiology, underlying problems associated with the type of infection, clinical features, diagnosis and therapy. An ample number of photographs is provided to illustrate major diagnostic features.

Brain Abscess

Etiology

Anaerobes found in the majority of cases — included are *Bacteroides fragilis* group, other *Bacteroides*, *Fusobacterium*, *Peptostreptococcus*. Less commonly — clostridia, *Actinomyces*, others.

Nonanaerobes — Viridans streptococci, microaerophilic streptococci, group A streptococci, pneumococci, *Staphylococcus aureus*, *Haemophilus aphrophilus*, *Actinobacillus*, others.

Underlying Problems

Sinusitis

Otitis media, mastoiditis

Oral, dental infection

Pulmonary infection

Congenital heart disease with
right-to-left shunt

Bacteremia, endocarditis

Clinical Features

Space-occupying mass

Severe headache (70% of patients)

Fever in only 50% of patients

Altered mental status (50 to 70%)

Nausea, vomiting (50%)

Focal neurologic defects

Seizures (30%)

Papilledema

Diagnosis

Lumbar puncture contraindicated

CT (with contrast) and/or MRI scan key study

Workup for underlying problem

Obtaining material for culture (e.g., via burr-hole) is highly desirable Optimum anaerobic transport important

Therapy

Excision or drainage not always required (especially when abscess not encapsulated and not close to ventricular wall)

Operation is mandatory when neurologic deficits are severe or progressive, especially when signs of brain stem compression are present

Antimicrobial therapy

Metronidazole plus penicillin G or ampicillin (or chloramphenicol) is best regimen for cases without bacteriologic information

Other drugs if indicated by clinical or bacteriologic data

Treat at least several weeks (judge by follow-up scans)

Measures to reduce intracranial pressure —

> Short-term treatment with corticosteroids

Anticonvulsants should be used initially

Anticoagulation contraindicated because of risk of hemorrhage

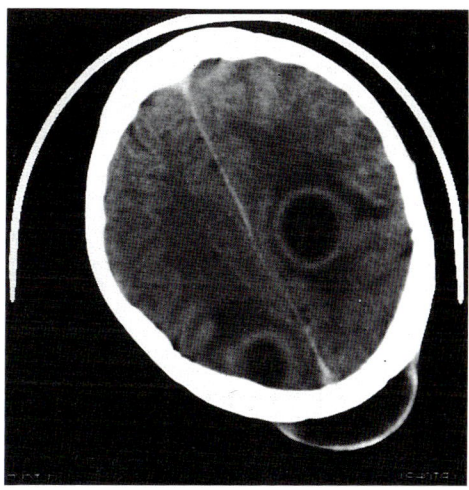

Figure 3.
CT scan, head showing two lesions consistent with brain abscess — ring enhancement (by contrast material) and surrounding edema.

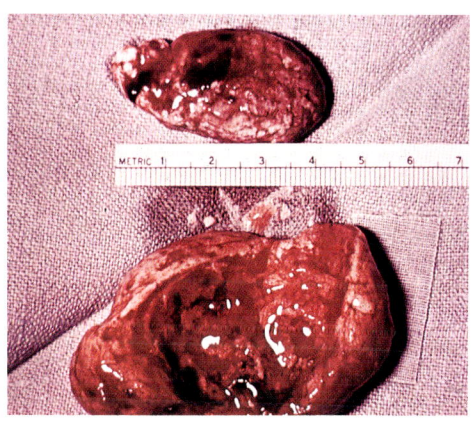

Figure 4.
Excised encapsulated brain abscesses.

Subdural Empyema

Etiology

35% of organisms are aerobic streptococci

17% are staphylococci

Haemophilus influenzae and facultative gram-negative rods are seen occasionally

12% are anaerobic or microaerophilic streptococci

Other anaerobes seen include *Bacteroides* (*B. fragilis* among them), *Fusobacterium*, clostridia, and *Actinomyces*

>25% of cultures show no growth; anaerobes, therefore, are probably more common

Underlying Problems

Sinusitis

Otitis media, mastoiditis

Meningitis

Neurosurgery, penetrating head trauma

Pleuropulmonary infection

Bacteremia

Infection of previous subdural hematoma

Dental infection

Pharyngotonsillar infection

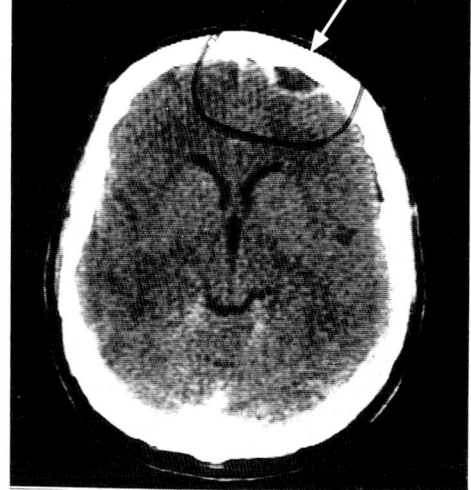

Clinical Features

Manifestations of underlying problem

Increasingly severe headache

Fever

Meningeal signs

Focal cerebral signs later

Increasing obtundation — to coma

III and VI cranial nerve palsies

Papilledema

Figure 5.
CT scan, head. Subdural empyema — dark area between brain and skull (arrow).

Diagnosis

CT scan is most sensitive and specific diagnostic procedure. Lesions typically unilateral and over convexity; 10% are adjacent to falx. Well-defined rim opacification is characteristic (chronic subdural hematoma does not show this).

Lumbar puncture should not be performed

Workup for underlying problem

Anaerobic transport of material for culture

Therapy

Surgical drainage

Penicillin or ampicillin plus metronidazole; chloramphenicol

Nafcillin plus rifampin or a third-generation cephalosporin for post-craniotomy cases

Other drugs if indicated by clinical or bacteriologic data

Spinal Epidural Abscess

This entity seldom involves anaerobes

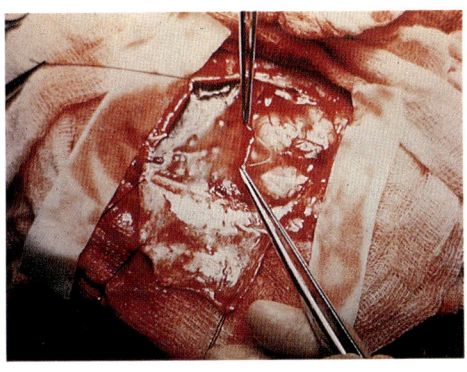

Figure 6.
Appearance of subdural empyema at surgery. Clamps are attached to incised dura revealing subdural pus.

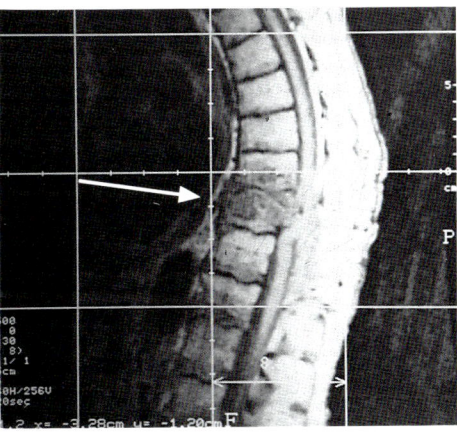

Figure 7.
Magnetic resonance image showing spinal epidural abscess (and vertebral osteomyelitis) (arrow).

Orofacial and Odontogenic Infection

Etiology

Primarily anaerobes and facultative streptococci

Anaerobes include *Bacteroides*, *Fusobacterium*, *Peptostreptococcus,* and *Actinomyces*

One study of hospitalized patients refractory to treatment found *B. fragilis* in nearly 30% of subjects

S. aureus may be found in osteomyelitis cases

Underlying Problems

Root canal infection

Periapical abscess

Dental caries

Periodontal disease, gingivitis

Dental surgery, trauma

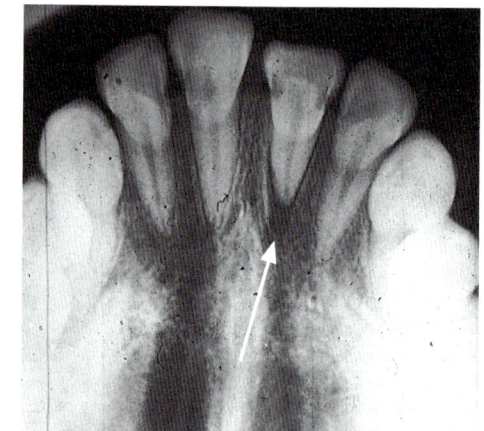

Figure 8.
Periapical abscess showing destruction of bone (black area) about the apex of one of the incisor teeth (arrow).

Clinical Features

Abscess formation, phlegmon

Pyogenic granuloma

Bacteremia at times

Osteomyelitis of jaw, osteitis

Fistulae

Diagnosis

Dental examination, radiographs

Blood cultures

Aerobic and anaerobic culture of abscess contents, other appropriate specimens

Good anaerobic transport of material to be cultured is important

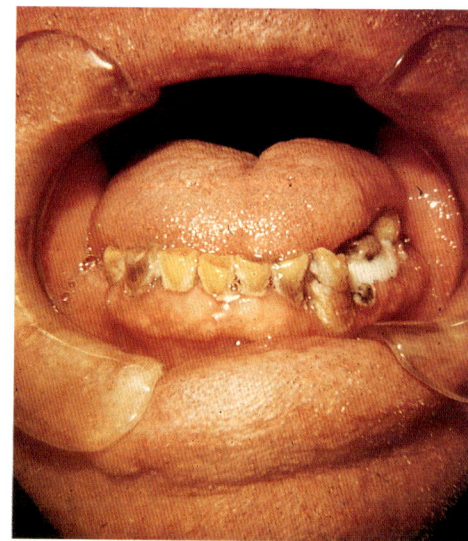

Figure 9.
Extensive periodontal disease with necrotizing gingivitis.

Therapy

Drainage, debridement

Antimicrobial therapy — Penicillin, ampicillin/sulbactam, amoxicillin/clavulanic acid, metronidazole (plus a penicillin), clindamycin, or tetracycline are all suitable, depending on the extent and severity of the process

Antistaphylococcal agents may be needed for some cases of osteomyelitis

Correction of underlying problem

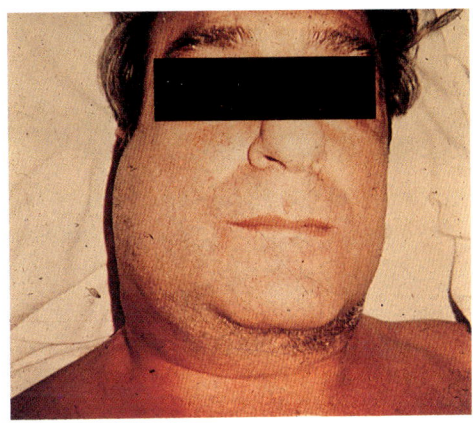

Figure 10.
Phlegmon, right jaw. Same patient as in Fig. 9.

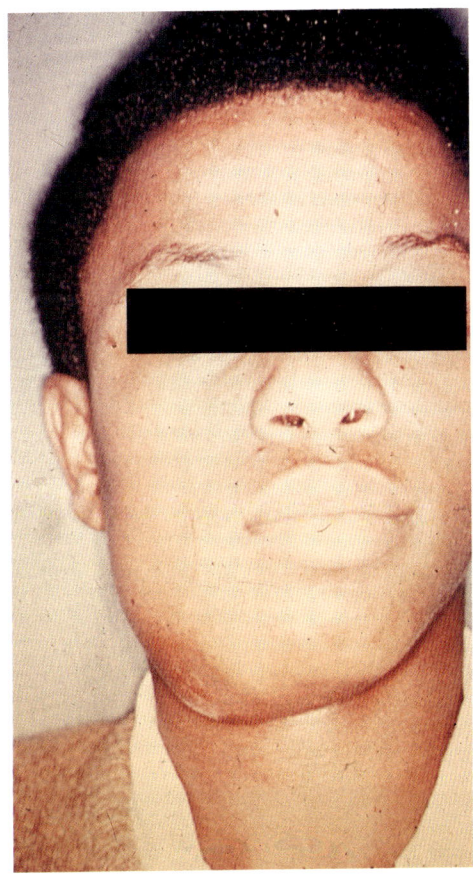

Figure 11.
Abscess of right jaw of dental origin.

Chronic Otitis Media

Etiology

Aerobes and facultatives — *Pseudomonas aeruginosa*, *S. aureus*, *Enterobacteriaceae*, *Streptococcus pneumoniae*, *H. influenzae*, *Moraxella (Branhamella) catarrhalis*, group A *Streptococcus*

Anaerobes — Probably present in well over 50% of cases with proper anaerobic transport and culture. Included are various *Bacteroides* species (*B. fragilis* group among them), *Peptostreptococcus*, *Fusobacterium*, *Veillonella*, clostridia, *Arachnia* and *Lactobacillus*.

Underlying or Associated Problems

Acute otitis media

Obstruction of Eustachian tube
(enlarged adenoids in children)

Mastoiditis

Cholesteatoma

Epidural or brain abscess, meningitis

Clinical Features

Persistent middle ear effusion

Persistent or recurrent drainage in
some patients

Foul odor to drainage common

Fever

Pain

Hearing loss

Perforation of drum

Diagnosis

History and physical examination

Aerobic and anaerobic culture of middle ear fluid (pus); anaerobic transport important

Radiographs or CT scans of mastoids

Therapy

Myringotomy

Adenoidectomy

Tympanostomy tube

Antimicrobial therapy appropriate for the organisms recovered. Ampicillin/ sulbactam, amoxicillin/clavulanic acid or clindamycin represent good choices in the absence of bacteriologic data.

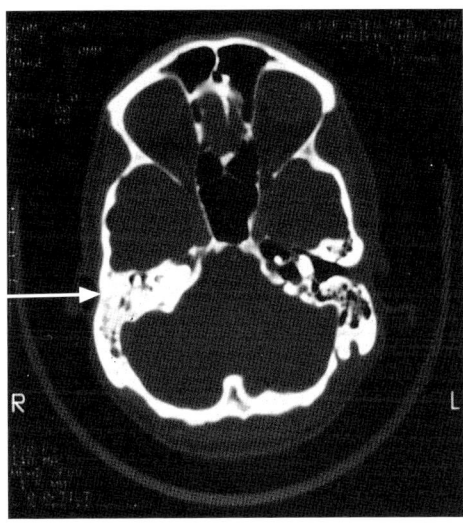

Figure 12.
CT scan showing mastoiditis, right (arrow). Note relative absence of right mastoid air cells as compared with left.

Chronic Sinusitis

Etiology

Anaerobes dominate the infecting flora — *Peptostreptococcus* spp., *Bacteroides* spp. (*B. fragilis* among them), *Fusobacterium*, *Eubacterium*, and *Actinomyces*

Nonanaerobes — various streptococci, *Haemophilus*, *S. aureus*, pneumococci, and *Enterobacteriaceae*

Underlying Problems

Acute sinusitis

Anatomic abnormalities — deviated septum, congenital choanal atresia

Foreign body (including indwelling nasal tubes)

Tumor

Allergy, polyps

Dental infection

Clinical Features

Persistent purulent nasal or postnasal discharge; may be foul-smelling

Facial pain, tenderness; headache

Fever

Cough, especially in children

Evidence of underlying disease or problem

Diagnosis

Transillumination (maxillary and frontal sinuses)

Radiologic and/or CT examination — opacity, mucosal thickening, or air-fluid level

Blood cultures

Antral puncture with aerobic and anaerobic culture; anaerobic transport of specimen important

Therapy

Nasal decongestants (avoid antihistamines as these may thicken fluid)

Sinus drainage and lavage, when feasible

Surgery — creation of artificial ostium for drainage, submucous resection

Antimicrobial therapy appropriate for the organisms recovered. For empiric therapy, ampicillin/sulbactam, amoxicillin/clavulanate or clindamycin represent good regimens.

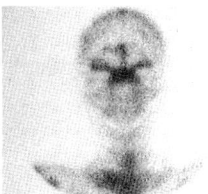

Figure 13.
Gallium isotope scan showing sinusitis involving multiple sinuses bilaterally.

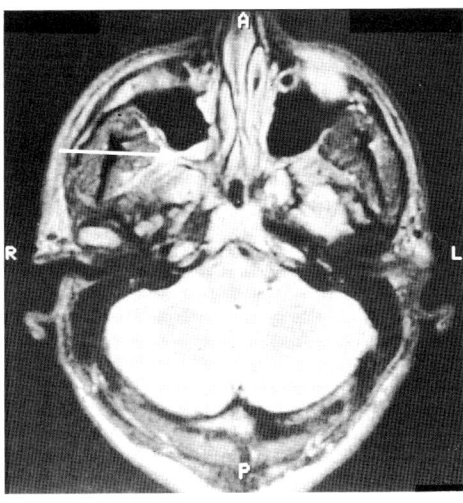

Figure 14.
CT scan showing right maxillary sinusitis with fluid level and some mucosal thickening (arrow).

Vincent's Angina and Postanginal Sepsis

Etiology

Fusobacterium necrophorum is the key pathogen, particularly when the complications result in sepsis. Spirochetes, and perhaps other organisms, may play a role.

Underlying Problems

None known

Clinical Features

Seen primarily in children and young adults

Pharyngitis or tonsillitis in most cases; may be so mild as to be ignored or even overlooked by patient — or may have resolved at the time of onset of sepsis

Tonsillar pseudomembrane may be present. Ulceration and necrosis seen on occasion. Foul discharge and consequent foul odor to breath

Septic internal jugular vein thrombophlebitis often present; local tenderness, swelling

Fever; rigors may be present, particularly with sepsis or septic thrombophlebitis

Pain, swelling at angle of jaw

Stiff neck, dysphagia; palpable cord, jugular vein

With sepsis, metastatic infection is common and may be devastating — septic pulmonary emboli leading to lung abscesses; empyema, mediastinitis, liver abscess, septic arthritis, and osteomyelitis are also especially common, but other organs may be involved

Peritonsillar abscess may occur

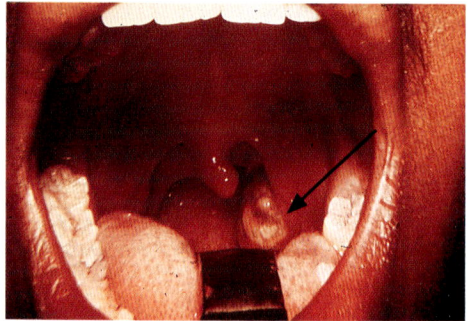

Figure 15.
Left peritonsillar abscess (right side of photo). Note pseudomembrane on tonsillar surface and bulging of left tonsil and soft palate, with medial displacement (arrow).

Diagnosis

Distinctive clinical picture

Anaerobic blood cultures

Methylene blue or Gram stain of pseudomembrane — fusiforms and spirochetes predominate

Gram stain and anaerobic culture of metastatic foci; anaerobic transport important

Ultrasound and CT of jugular vein

Therapy

Penicillin G is drug of choice; use high dosage (15 to 20 million units/day in adults). Metronidazole is a good alternative

Erythromycin is *not* effective (may have been used initially for pharyngotonsillitis)

Drainage of local or metastatic abscesses, empyema, as indicated

Ligation or excision of internal jugular vein may be required; anticoagulation could be risky (proximity of carotid artery, which may be eroded)

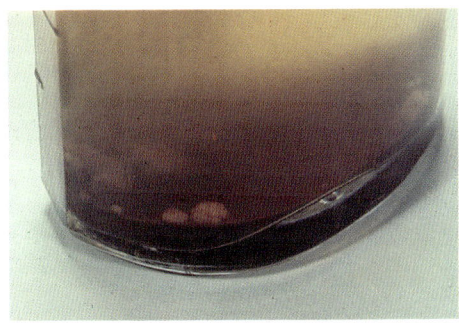

Figure 16.
Fluffy colonies of Fusobacterium necrophorum *at the bottom of a blood culture bottle.*

Neck Space Infections (Ludwig's Angina)

Etiology

Anaerobes predominate — *Peptostreptococcus*, *Bacteroides* (*B. fragilis* group among them), and *Fusobacterium* primarily

Nonanaerobes — dominated by viridans group streptococci; group A streptococci may also be seen

Underlying Problems

Dental infections, surgery

Tonsillar infections, peritonsillar abscess

Pharyngitis

Sinusitis

Sepsis

Malignancy

Trauma

Head and neck surgery

Clinical Features

Uncommon, but often life-threatening infections

Presence of underlying problem

Features vary with particular neck space involved and underlying problem. *Ludwig's angina* is a cellulitis (rather than abscess) that involves the major anterior neck compartment, the submandibular space (itself divided into submaxillary and sublingual spaces). The classic features of this process are massive, diffuse, bilateral board-like cellulitis, bulging of the floor of the mouth with elevation of tongue, dysphagia, drooling, dysphonia, pain, trismus, and, at times, respiratory distress. The tissue is gangrenous and

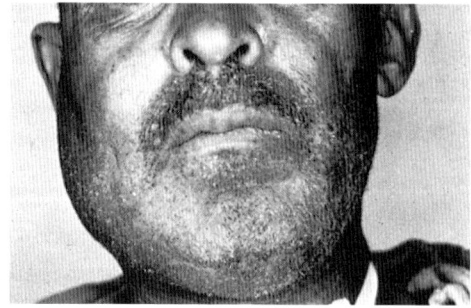

Figure 17.
Ludwig's angina; note diffuse swelling under chin.

the exudate serosanguinous, with little or no pus. The risks are asphyxiation due to edema of the neck and glottis and aspiration pneumonia.

Fever, systemic toxicity

Diagnosis

Distinctive clinical picture, including evidence of underlying disease

Dental examination, radiographs

Radiographs, CT scan of paranasal sinuses

Lateral neck radiograph (soft tissue swelling, gas)

CT scan of neck

Gram stain, aerobic and anaerobic culture of material obtained at surgery or by aspiration; optimum anaerobic transport important

Blood cultures — aerobic and anaerobic

Surgical consultation should always be obtained when deep neck space infections are suspected

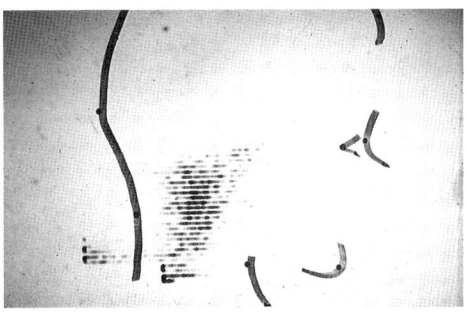

Figure 18.
Gallium isotope scan showing retropharyngeal abscess in young girl with Vincent's angina.

Therapy

Maintenance of airway; early tracheostomy or intubation, when necessary

Decompression of soft tissues

Surgical debridement, drainage

Intensive antimicrobial therapy — useful regimens would include penicillin plus metronidazole, ampicillin/sulbactam, ticarcillin/clavulanic acid, imipenem

Bacteremia

Etiology

The anaerobes recovered most commonly from bacteremia are *B. fragilis* and other members of this group, *Clostridium perfringens* and other clostridia, and anaerobic cocci. To some extent, this reflects the ease with which these anaerobes are recovered from blood. The incidence of anaerobes in positive blood cultures is 5 to 15% in various studies.

Portals of Entry

Gastrointestinal tract most common

Female genital tract

Decubitus ulcer

Infected extremity ulcers related to vascular disease

Dental manipulation

Upper respiratory tract

Lower respiratory tract — uncommon

Clinical Features

Fever, chills

Hypotension, shock, coagulopathy

Hyperbilirubinemia

Thrombophlebitis (suppurative at times)

Metastatic suppurative infection

} Particularly common in anaerobic bacteremia

Clostridium septicum bacteremia associated with malignancy of colon, especially cecum

Propionibacterium acnes bacteremia associated with foreign bodies

Diagnosis

Anaerobic blood cultures (hold for 2 weeks)

Gram stain and anaerobic (and aerobic) culture of underlying portal of entry and metastatic sites; good anaerobic transport important

In the case of *P. acnes*, which is seen commonly as a contaminant (from normal skin flora), careful skin cleansing, repeated cultures, and perhaps quantitative cultures, will help to decide the significance of an isolate.

Therapy

Early appropriate antimicrobial therapy — according to organism isolated. For empiric therapy, good choices include metronidazole plus penicillin, ampicillin/sulbactam, imipenem or chloramphenicol.

Usual supportive therapy (for shock, etc.)

Ligation or anticoagulation may be required for thrombophlebitis

Figure 19.
Blood culture positive for C. perfringens.
Note hemolysis and gas production.

Endocarditis

Etiology

Obligate anaerobes uncommon cause (1 to 15%), but perhaps increasing

Most common anaerobes include *B. fragilis*, *Peptostreptococcus*, *F. necrophorum*, *Clostridium* spp., and *Propionibacterium*

Underlying Problems

Valvular heart disease in about 50% of patients

Implanted prosthetic material (heart valve, vascular graft, or intravascular prosthesis)

Portal of entry — see **Bacteremia** section, page 32

Clinical Features

Course usually subacute

B. fragilis commonly leads to large vegetations and embolization

Thrombophlebitis

Congestive heart failure in one-fourth to one-third of patients

Valvular destruction more frequent than with viridans streptococci

Rupture of chordae tendineae

Cardiogenic or septic shock

Arrhythmias

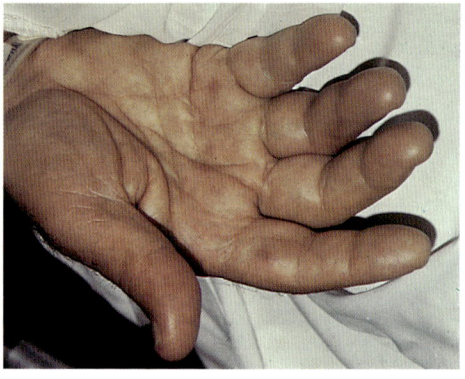

Figure 20.
Janeway lesions of acute bacterial endocarditis.

Diagnosis

Anaerobic blood cultures

Gram stain and culture of embolic sites, when feasible

Two-dimensional echocardiogram

Electrocardiogram (re: myocarditis and myocardial abscess)

Therapy

Bactericidal drug important

Choice of drug depends on organism. Metronidazole would be excellent against most anaerobes. Penicillin G would be preferable for *P. acnes* and some anaerobic or microaerophilic cocci.

Treat for at least 3-4 weeks

Follow-up blood cultures important

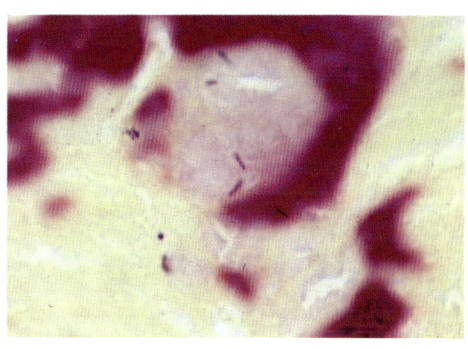

Figure 21.
Lactobacillus *endocarditis. Gram stain of heart valve showing thin, regular gram-positive bacilli embedded in tissue.*

Lung Abscess, Aspiration Pneumonia, Empyema

Etiology

Oral and gastric flora. Anaerobes found in 90% of pulmonary infections and two-thirds of empyemas

1. Community acquired. Endogenous flora — viridans group streptococci and various anaerobes including *Bacteroides* (7% of patients have *B. fragilis* group), *Fusobacterium*, *Peptostreptococcus* and, less commonly, *Clostridium*, *Eubacterium*, *Actinomyces*, and others

2. Hospital acquired. Endogenous flora, as outlined above, plus nosocomial pathogens colonizing the upper airways (*S. aureus*, *Enterobacteriaceae*, and *P. aeruginosa*, primarily)

Underlying Problems

Aspiration (altered consciousness, mechanical disruption of defense barriers, dysphagia)

Gingivitis, periodontal disease

Pulmonary neoplasm, foreign body

Pulmonary embolism; septic emboli

Bronchiectasis

Intra-abdominal sepsis

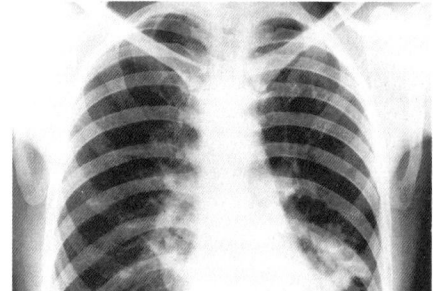

Figure 22.
PA chest film of patient with bronchiectasis. Note circular infiltrates, some filled with air, at both bases adjacent to heart.

Clinical Features

Fever, leukocytosis; no rigors

Subacute presentation common; may be acute

Weight loss, anemia

Putrid sputum or empyema fluid (half of patients)

Involvement of dependent segment(s)

Initial phase is pneumonitis; later, there may be multiple small excavations (necrotizing pneumonia), frank abscess formation (most often solitary) and/or empyema

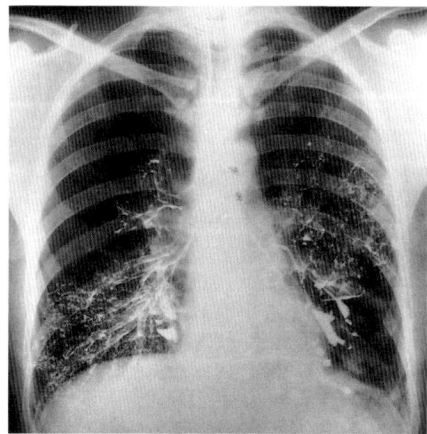

Figure 23.
Bronchogram demonstrating bilateral saccular bronchiectasis adjacent to heart borders.

Necrotizing pneumonia more serious than other types

Predisposition to aspiration or other typical underlying problem

Diagnosis

Chest radiograph, CT

Typical location (dependent segment)

Foul sputum or empyema fluid (not always present)

Typical history — common background factors, subacute course

Blood culture yield very low

Gram stain, aerobic and anaerobic culture of pleural fluid, transtracheal aspirate or other specimen free of oropharyngeal flora (sputum is *not* suitable)

Therapy

Drainage of empyema. This usually requires one or more chest tubes and sometimes an open drainage procedure with rib resection and breakdown of pleural space loculations.

Postural drainage, cupping may be useful in patients with large abscesses.

Surgery is rarely indicated, and usually contraindicated, for lung abscess.

Antimicrobial therapy —

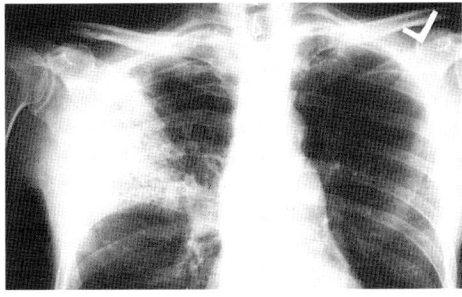

Figure 24.
PA chest film. Necrotizing aspiration pneumonia involving posterior segment of right upper lobe. Note multiple small excavations and elevated horizontal fissure (due to inflammatory edema of bronchus).

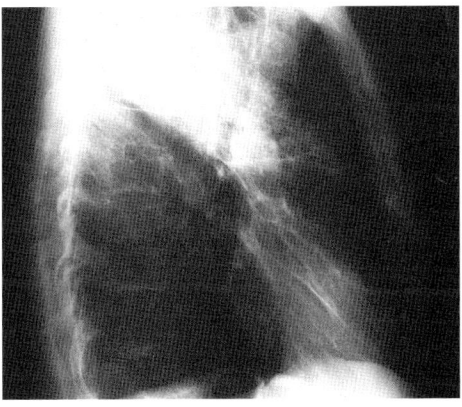

Figure 25.
Lateral chest film, of patient seen in Figure 24, showing posterior location of infiltrate. The posterior segments of the upper lobes are one of the two most common sites for aspiration.

1. Community-acquired cases — there are a number of choices, depending on the severity of illness. Clindamycin, cefoxitin, metronidazole plus penicillin, amoxicillin/clavulanate, and ampicillin/sulbactam are all suitable.

2. Hospital-acquired — depends on severity of illness and aerobic culture results. When *P. aeruginosa* is present, two agents active against the infecting strain should be employed (to be selected from drugs such as ceftazidime, imipenem, cefsulodin, piperacillin, ticarcillin/clavulanate and aminoglycosides). This regimen will cover most *Enterobacteriaceae* well, but not necessarily *S. aureus* and the anaerobes.

3. Antimicrobial therapy should be prolonged in order to reduce the risk of relapse.

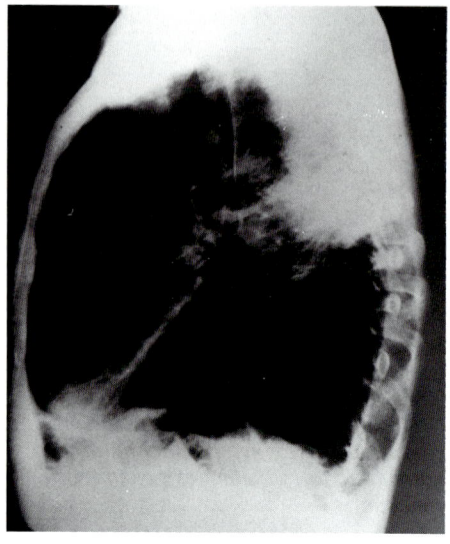

Figure 26.
Lateral chest film of patient with aspiration pneumonia involving superior segment of lower lobe, the second most common site of aspiration.

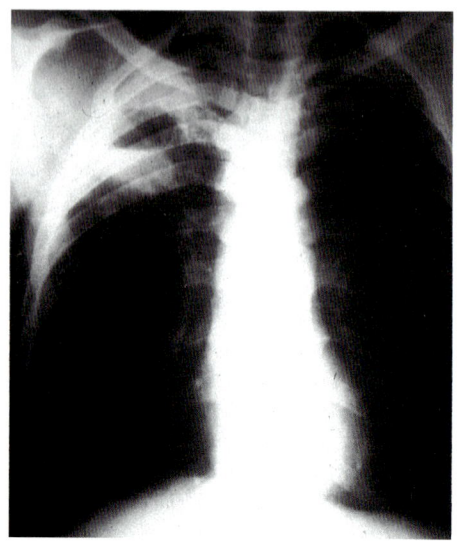

Figure 27.
Lung abscess, right upper lobe. Note air-fluid level.

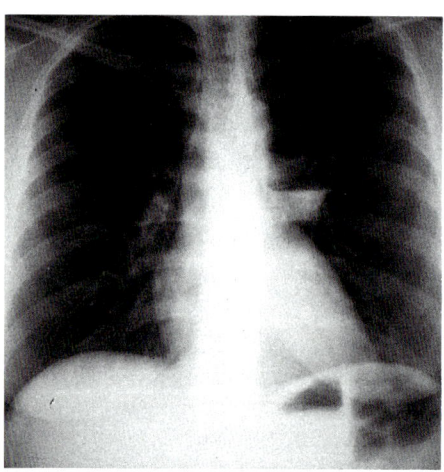

Figure 28.
Lung abscess, left mid-lung field.

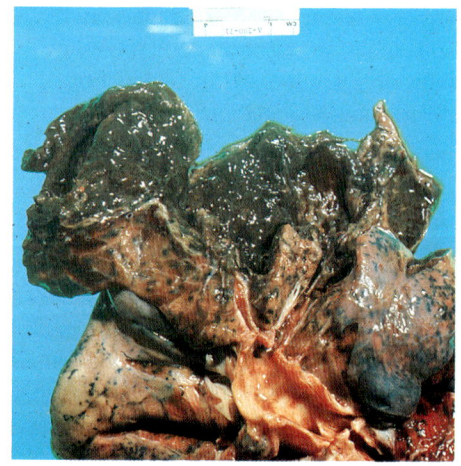

Figure 29.
Autopsy specimen of large, multiloculated right upper lobe abscess. (Patient died of other causes.)

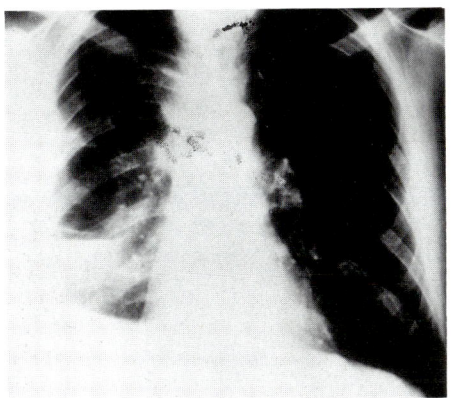

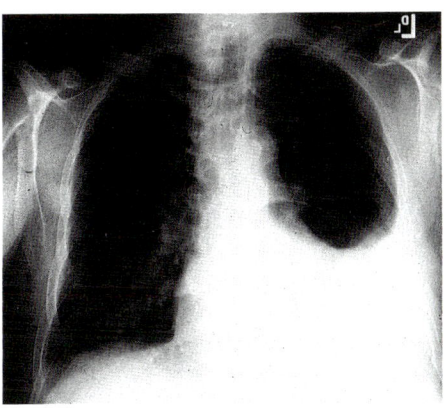

Figure 30.
PA chest film of patient with very
large lung abscess plus empyema (on
right).

Figure 31.
PA chest film showing large left
pleural effusion.

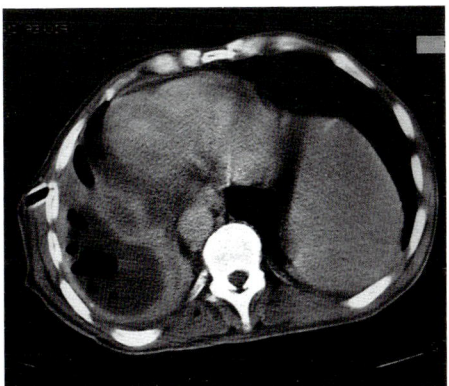

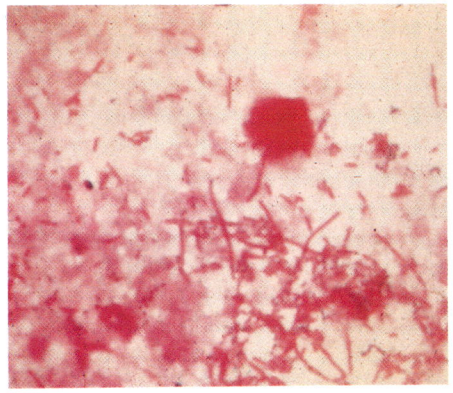

Figure 32.
CT scan of chest showing
empyema with gas production. Two
definite gas-fluid levels can be seen.

Figure 33.
Gram stain of transtracheal aspirate
from patient with anaerobic pulmonary
infection. Predominant organism is
filamentous, pleomorphic, gram-negative
*bacillus (*F. necrophorum*). Also seen in*
lower left corner are small gram-negative
*coccobacilli (*Bacteroides
melaninogenicus*) and in the lower right*
corner a pair of gram-positive cocci
*(*Peptostreptococcus anaerobius*).*

Liver Abscess

Etiology

Probably over half of liver abscesses would yield anaerobic bacteria if cultured properly

Anaerobes — in approximate order of frequency, anaerobic streptococci, microaerophilic streptococci, *B. fragilis* group, other *Bacteroides*, *F. necrophorum*, other fusobacteria, *Clostridium*, and *Actinomyces*

Nonanaerobes — viridans streptococci, enterococci, enteric gram-negative bacilli, group A *Streptococcus*, and *S. aureus* (the latter in association with microabscesses)

Underlying Problems

Biliary tract infection, especially cholangitis

Hematogenous seeding (systemic bacteremia)

Seeding from portal venous bacteremia — related to diverticulitis, peritonitis, intra-abdominal and perirectal abscess, appendicitis and omphalitis

Direct extension of contiguous infection — penetrating peptic ulcer, cholecystitis, pancreatitis, perihepatic abscess

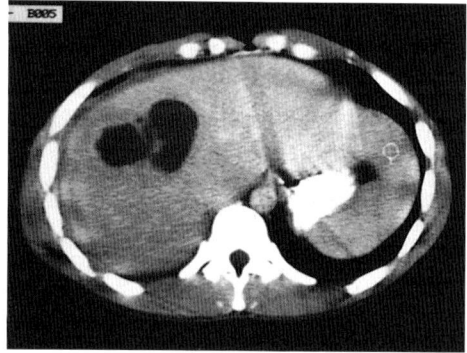

Figure 34.
CT scan, abdomen, showing large, multiloculated abscess in liver.

Clinical Features

Fever, leukocytosis

Right upper quadrant or other abdominal pain and tenderness

Hepatomegaly

Weight loss, anemia

Pleurisy, pain referred to right shoulder

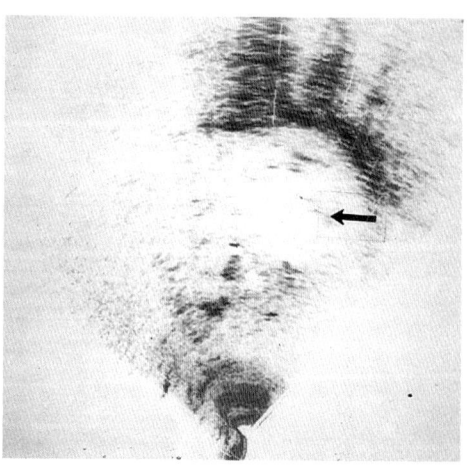

Figure 35.
Ultrasound of abdomen showing liver abscess (arrow).

Diagnosis

Elevated alkaline phosphatase with other liver function tests normal or relatively normal

CT, ultrasound

Liver scan

Radiography — elevated right diaphragm, pleural effusion, right basilar pulmonary infiltrates

Blood cultures (anaerobic) commonly positive

Negative amebic serology

Therapy

Percutaneous drainage is often effective as an alternative to surgical drainage; occasional cures without any drainage procedure have been reported

Empiric therapy, pending bacteriologic data, might be metronidazole plus ampicillin and an aminoglycoside, ampicillin/sulbactam plus an aminoglycoside, imipenem, or ticarcillin/clavulanate

Therapy should be prolonged, particularly with multiple abscesses

Figure 36.
Liver scan showing abscess (upper picture is anteroposterior view, lower is lateral view).

Biliary Tract Infection

Etiology

Anaerobes are found in 10 to 25% of these infections

Aerobic and facultative bacteria — *Enterobacteriaceae* (especially *E. coli*), *Pseudomonas*, enterococci, viridans streptococci and, rarely, *S. aureus*

Anaerobes — *B. fragilis*, *C. perfringens*, and *Peptostreptococcus* primarily

Underlying Problems

Bile stasis — calculi, strictures, tumors

Acute inflammation

Calculi

Clinical Features

Epigastric or right upper quadrant pain

Nausea, vomiting

Fever, often low-grade in elderly

Shaking chills

Jaundice may be present

Diagnosis

Blood cultures (aerobic and anaerobic)

Liver function tests often abnormal

Gram stain and aerobic and anaerobic
culture of bile

Ultrasound

Hepatobiliary scanning (e.g., HIDA
scan)

Therapy

Surgery for obstruction, empyema of
gallbladder, tumor

Antimicrobial therapy, pending
bacteriologic data — cefoxitin,
ampicillin plus metronidazole, and
ampicillin/sulbactam are reasonable
empiric choices

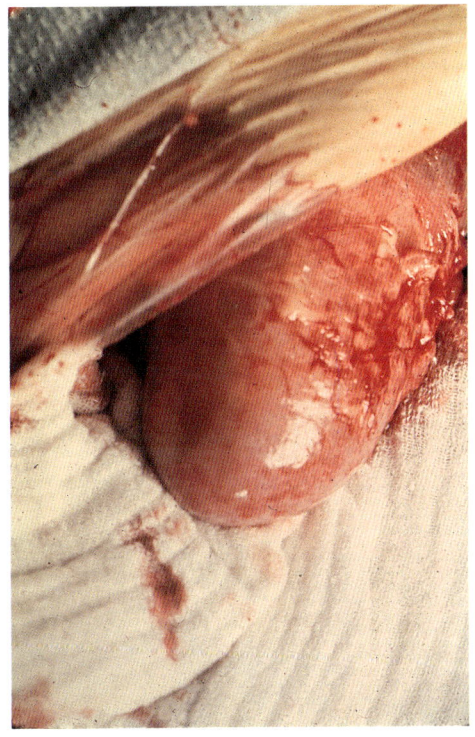

Figure 37.
Empyema of the gallbladder exposed at
time of surgery.

Peritonitis, Intra-Abdominal Abscess

Etiology

Primary peritonitis seldom involves anaerobes and will not be discussed further.

Secondary peritonitis and intra-abdominal abscess involve primarily elements of the gastrointestinal flora.

Various studies show approximately equal numbers of anaerobes and nonanaerobes; it is likely that studies with optimum anaerobic transport and culture techniques would show a predominance of anaerobes since these organisms greatly outnumber nonanaerobes in the normal bowel flora.

Anaerobes — *B. fragilis* group predominantly, other *Bacteroides*, *Clostridium* spp. *Peptostreptococcus* spp., and various groups of gram-positive non-spore-forming rods.

Nonanaerobes — dominated by *Escherichia coli*; also seen are other *Enterobacteriaceae*, streptococci of various types (including enterococci), *S. aureus*, and *Pseudomonas* spp.

Underlying Problems

Gastrointestinal surgery

Postoperative anastomotic leak

Perforated or gangrenous appendicitis

Perforated peptic ulcer

Diverticulitis

Inflammatory bowel disease

Traumatic or neoplastic perforation of the gastrointestinal tract

Intestinal ischemia or infarction

Biliary tract disease

Pancreatitis

Clinical Features

Varying degrees and locations of abdominal pain, tenderness, guarding

Nausea, vomiting may be seen

Peritoneal signs often masked in postoperative patient because of incisional pain and narcotics

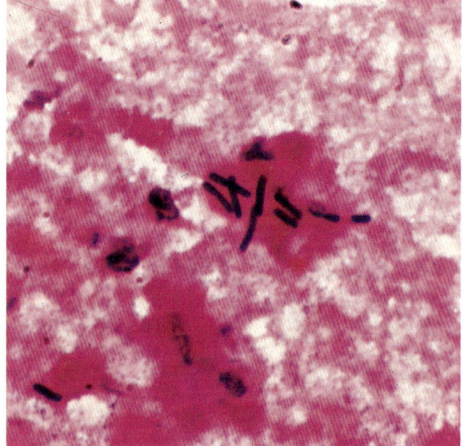

Figure 38.
Gram stain of peritoneal fluid from patient with peritonitis. The only organisms present have morphology highly suggestive of C. perfringens — *large, boxcar-like, gram-positive rods without apparent spores (* C. perfringens *does not sporulate readily).*

Fever, sometimes chills

Ileus

Evidence of underlying problem

Chest findings common in subphrenic abscess

Rectal, pelvic tenderness, fluctuance

Diagnosis

Blood cultures (aerobic and anaerobic)

Gram stain, aerobic and anaerobic culture of pus or fluid from abdominal cavity; optimum anaerobic transport important

Ultrasound and CT scans (latter is the optimal procedure)

Gallium or indium scans

Upright and flat abdominal films, chest film

Barium contrast studies useful on occasion

Diagnostic aspiration of abdominal cavity

Laparotomy may be required

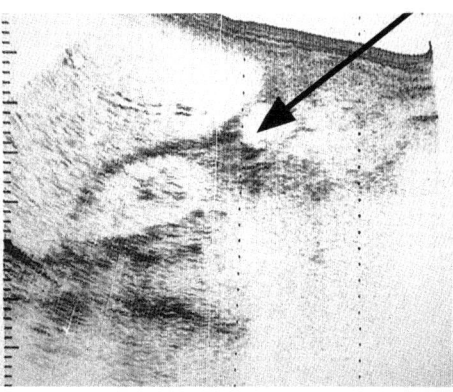

Figure 39.
Ultrasound of abdomen showing pericecal abscess (arrow).

Therapy

Drainage (percutaneous or surgical) and debridement paramount

Repair of anastomotic or other leaks

Antimicrobial therapy — clindamycin plus gentamicin and cefoxitin are two widely used regimens that have given generally good results despite significant resistance among strains of the *B. fragilis* group and clostridia other than *C. perfringens*. Many other regimens are also recommended, including metronidazole plus penicillin and an aminoglycoside, imipenem, ticarcillin/clavulanate, and ampicillin/sulbactam.

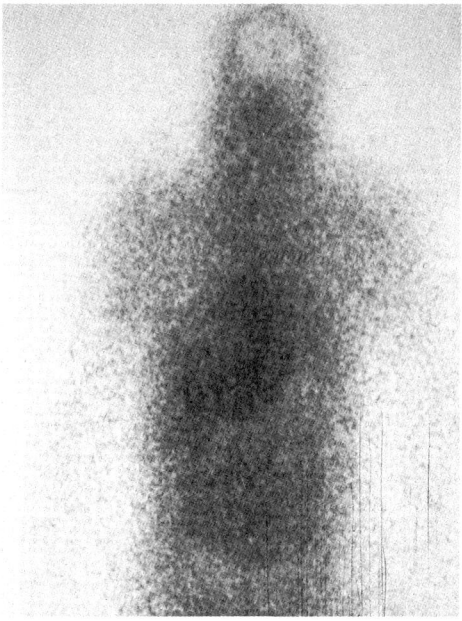

Figure 40.
Gallium isotope scan showing diffuse peritonitis.

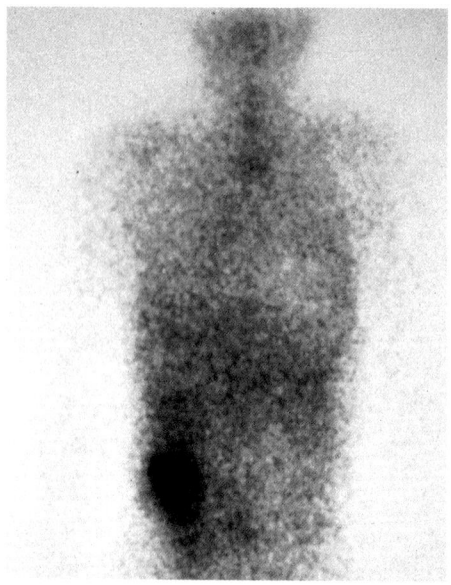

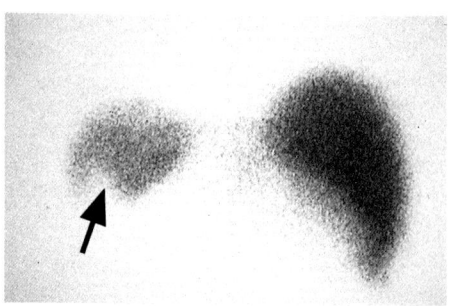

Figure 41.
Gallium isotope scan showing right
lower quadrant abscess.

Figure 42.
Liver-spleen scan (posterior view)
showing an abscess in lower pole of
spleen (arrow).

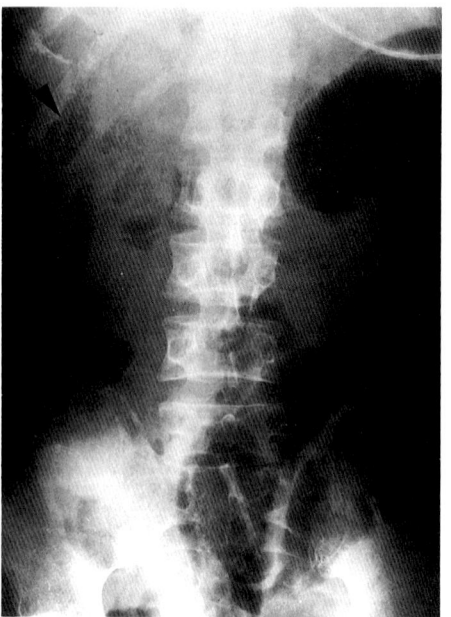

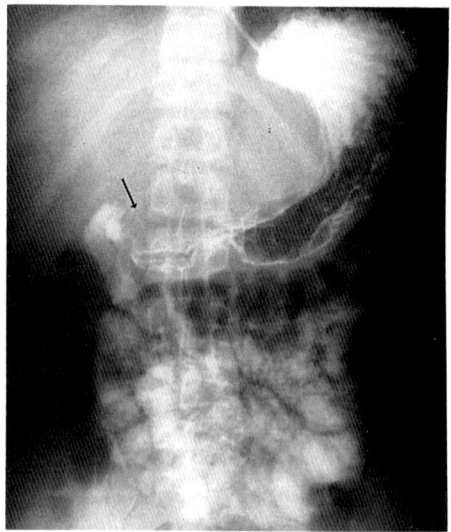

Figure 43.
Flat plate of abdomen showing
extraintestinal gas in an abscess (arrow,
right upper quadrant).

Figure 44.
Barium swallow showing an
anastomotic leak (arrow) and large
abscess compressing lesser curvature
of stomach and small bowel.

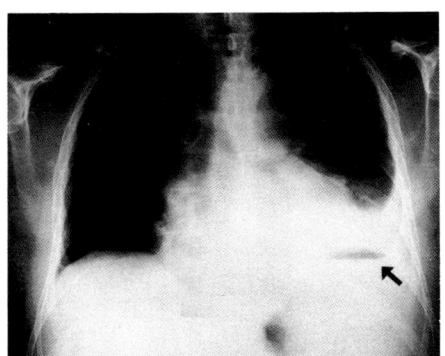

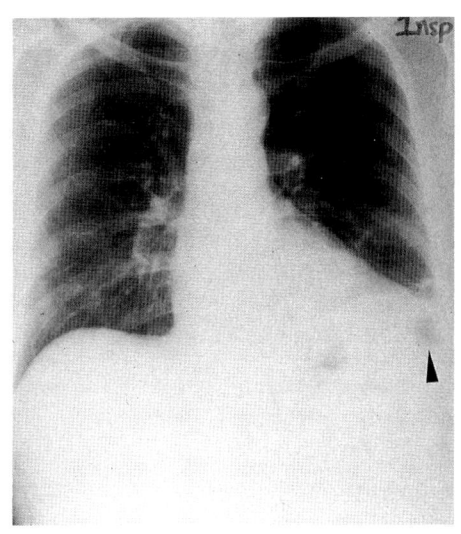

Figure 45.
Left subphrenic abscess. Note elevation
of left diaphragm, left pleural effusion
and gas-fluid level in abscess (arrow).

Figure 46.
Left subphrenic abscess. Elevated
left diaphragm, blunting left costophrenic
angle (fluid) and small gas-fluid level in
abscess (arrow).

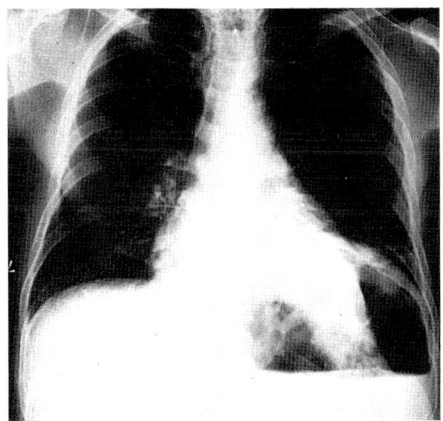

Figure 47.
Left subphrenic abscess. Findings
similar to those in Fig. 46, except for
huge gas-fluid level in abscess in this
patient.

Appendicitis

Etiology

Polymicrobial, with major predominance of anaerobes (average of 10.2 organisms per specimen from patients with gangrenous or perforated appendicitis; this includes average of 7.5 anaerobes and 2.7 nonanaerobes).

Predominant anaerobes — *B. fragilis* group (especially *B. fragilis* and *B. thetaiotaomicron*) and *Bilophila wadsworthia* (a newly described bile-resistant anaerobic gram-negative rod). Other anaerobes include other *Bacteroides* spp., *Fusobacterium*, anaerobic non-spore-forming gram-positive rods, anaerobic and microaerophilic streptococci, and clostridia.

Predominant nonanaerobes — *E. coli*, viridans streptococci and other streptococci (including enterococci), *Pseudomonas*, and other *Enterobacteriaceae*.

Underlying Problems

Obstruction due to a fecalith or, less commonly, hyperplastic lymphoid tissue leads to increased intraluminal pressure, mucosal necrosis and bacterial invasion.

Clinical Features

More common in younger age group but occurs at any age

Initial periumbilical pain, sometimes colicky

Subsequent right lower quadrant pain, maximal at McBurney's point

Right lower quadrant tenderness, guarding, rebound tenderness

Nausea

Atypical presentations in one-fourth of patients — young children, the elderly, in pregnancy, and in the case of retrocecal or pelvic location of the appendix

Pyuria or hematuria may occur

Moderate leukocytosis is typical — up to 20,000/mm^3

Diagnosis

Typical clinical picture

Rule out mesenteric lymphadenitis (in children, especially); disease of ovary, cecum; regional enteritis

Radionuclide scans, ultrasound and CT may be useful in detecting complicating abscess

Laparoscopy, laparotomy

Therapy

Early surgery — appendectomy and drainage as indicated

Antibiotics are useful in the case of gangrenous or perforated appendicitis and when there are complications. With serious complications, such as generalized peritonitis or bacteremia, regimens of choice would include ampicillin/sulbactam, metronidazole plus ampicillin, imipenem, and chloramphenicol. For other cases, suitable regimens include cefoxitin, ceftizoxime, and clindamycin plus gentamicin.

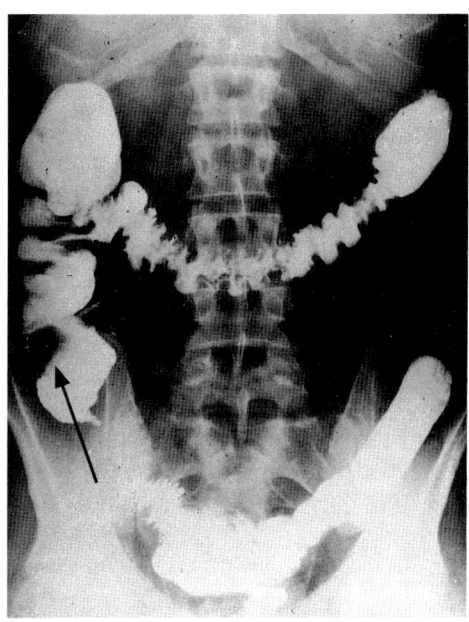

Figure 48.
Appendiceal abscess producing
constant filling defect (arrow) in lateral
wall of cecum.

Diverticulitis

Etiology

The bacteriology of diverticulitis has not been worked out in detail as it has been with appendicitis, but it is virtually certain that the flora of this condition would be similar to that of appendicitis (see section on Appendicitis, page 48, for details).

Underlying Problems

Diverticulosis of the colon

Obstruction is a factor, but the inflammation of diverticulitis results primarily from perforation (usually of a single diverticulum)

Clinical Features

Diverticulosis is uncommon under the age of 40 and increases in frequency with age

Location of pain depends on site and extent of diverticulitis

Pain is aggravated by increased intra-abdominal pressure as with defecation

Constipation is usual, but there may be diarrhea

Nausea and vomiting are not uncommon

Rectal bleeding occurs in one-fourth of patients; occasionally, this is massive

A tender mass may be felt on abdominal examination

Diagnosis

Barium enema (must be done carefully) may demonstrate the characteristic finding of an abscess cavity communicating with the colon lumen, as well as other diverticula. There may be free air due to perforation, evidence of obstruction, and fistulae to adjacent organs.

Sigmoidoscopic findings of importance include limited mobility of a segment of bowel that is normally freely moveable, abnormally sharp angulation in the region of the rectosigmoid or higher, narrowing of the bowel lumen, and an extraluminal mass.

With complicating abscess formation, radionuclide scans, CT scan and ultrasound are useful.

Therapy

Without complications, treatment of diverticulitis is medical. Suitable regimens include cefoxitin or clindamycin plus gentamicin. In the presence of significant complications the regimens of choice would include ampicillin/sulbactam, metronidazole plus ampicillin, imipenem, or chloramphenicol. The addition of an aminoglycoside may be considered.

Approximately 10% of patients require surgery. Indications for elective surgery are recurrent episodes of diverticulitis, progressive stenosis, palpable mass that does not regress with antimicrobial therapy, refractory dysuria (colovesical fistula), fistulae of other types, recurrent bleeding of significance, and failure to rule out carcinoma.

Early surgery is required in the case of serious complications such as obstruction or significant perforation.

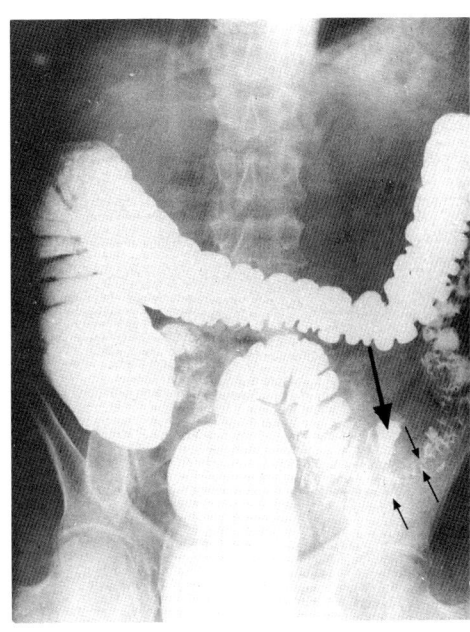

Figure 49.
Diverticulitis. Multiple sigmoid diverticula (arrows), one of which has perforated to produce (large arrow) a pericolonic abscess.

Perinephric Abscess

Etiology

Most perinephric abscesses are the result of preceding urinary tract infection. In these cases *Enterobacteriaceae* and *Pseudomonas* spp. are the primary infecting agents, but streptococci (primarily enterococci) may be involved.

Perinephric abscess of hematogenous origin secondary to skin or soft tissue infection may involve *Staphylococcus aureus*.

Anaerobic bacteria are important in the presence of obstruction leading to urinary extravasation, in relation to spread of infection from colon perforation, in renal transplant patients and in association with renal tumors. Anaerobes involved include *Bacteroides* (especially the *B. fragilis* group), *Fusobacterium* spp., *Clostridium* spp., anaerobic cocci, and *Actinomyces*.

Underlying Problems

Renal infection

Obstruction of the urinary tract and urinary extravasation

Extension from neighboring organs

Hematogenous spread

Diabetes mellitus and corticosteroid therapy predispose to this infection

Clinical Features

Onset is typically insidious

Fever, pain and flank tenderness are the most common features; fever is usually low-grade, but chills may occur. Pain most common in flank and costovertebral angle.

Palpable mass or bulging in the flank occurs in about half of patients

Psoas muscle spasm and spasm of paravertebral muscles are seen occasionally

Nausea and vomiting

Weight loss

Diagnosis

Typical clinical picture

Urine culture. If anaerobes are suspected, voided urine cannot be used; percutaneous bladder aspiration is suitable. Anaerobic transport should be used.

Aerobic and anaerobic blood cultures

Abdominal roentgenograms may demonstrate abnormal psoas outline, vertebral scoliosis or obliteration of renal outline by a mass. Gas bubbles may be seen occasionally.

Chest film may show elevated hemidiaphragm, pleural effusion, or basilar infiltrate

Radionuclide scans, ultrasound, and CT scans are important diagnostic aids. It may be possible to obtain material for Gram stain and culture by percutaneous needle aspiration of an abscess under ultrasound or CT guidance.

Intravenous pyelography may demonstrate extravasation of contrast material or fistula formation

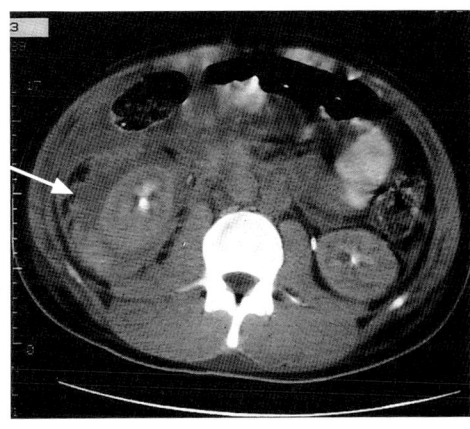

Figure 50.
CT scan of abdomen showing large
right perinephric abscess (arrow).

Therapy

Drainage is crucial. Surgery is usually indicated, but drainage under ultrasound or CT guidance may be adequate if there is no underlying problem.

Antimicrobial therapy is guided by Gram stain and culture results. If empiric therapy is required, suitable regimens would include ticarcillin/clavulanic acid, ampicillin/sulbactam, metronidazole plus ampicillin, or imipenem, all supplemented with an aminoglycoside. Vancomycin or another antistaphylococcal agent may be added, if indicated, to the metronidazole plus ampicillin regimen. Cefoxitin plus an aminoglycoside is another option.

Endometritis

Etiology

Anaerobes — more prevalent. Included are various *Bacteroides* (especially *B. bivius*, and *B. disiens*, the pigmented *Bacteroides* and *Porphyromonas* and the *B. fragilis* group), *Peptostreptococcus* spp., and clostridia. *Actinomyces* and *Eubacterium* spp., especially *Eubacterium nodatum*, are seen in association with intrauterine device (IUD) insertion.

Nonanaerobes — *E. coli* and other *Enterobacteriaceae*, groups A, B and D streptococci, and *Mycoplasma hominis*. In the case of endometritis related to sexual transmission, gonococci, *Chlamydia* and, perhaps, *Mycoplasma* or *Ureaplasma* may be involved.

Underlying Problems

Sexual transmission

Vaginal delivery

Delivery by Cesarean section

Complication of dilatation and curettage

Complication of IUD insertion

Risk factors include younger age, lower socioeconomic status, prolonged labor and premature rupture of membranes

Clinical Features

Fever, tachycardia

Lower abdominal pain

Uterine tenderness

Foul-smelling lochia

Signs and symptoms usually present within 3 days following delivery

Signs of peritonitis may be present

Endometritis associated with pelvic inflammatory disease (PID) may be accompanied by irregular menses

Diagnosis

Typical clinical picture

Presence of underlying problems

Aerobic and anaerobic blood cultures

Endometrial material for Gram stain and aerobic and anaerobic culture is best obtained by endometrial biopsy using a sheathed device such as the Pipelle®

Therapy

Most antimicrobial therapy of endometritis is empiric. Suitable regimens include ampicillin/sulbactam, metronidazole plus ampicillin, cefoxitin, ceftizoxime, or clindamycin plus gentamicin.

Pelvic Inflammatory Disease, Salpingitis, Tubo-Ovarian Abscess

Etiology

Most infections are polymicrobial. The flora is made up of the endogenous vaginal and cervical flora plus, in sexually transmitted infections, *Neisseria gonorrhoeae*, *Chlamydia trachomatis*, and *Mycoplasma*.

The most common **nonanaerobes** are *E. coli* and various streptococci.

Among the **anaerobes**, *Peptostreptococcus* spp. predominate with *Bacteroides* spp. (especially *B. bivius*, *B. disiens* and *B. fragilis*) also commonly encountered. In cases associated with an IUD, *Actinomyces* spp. and *Eubacterium nodatum* may be encountered.

Underlying Problems

Sexually acquired

Complication of dilatation and curettage

Complication of IUD insertion

May follow childbirth

May occur as a postoperative infection

Clinical Features

Abdominal pain is the most common presenting complaint; usually present for 1 to 2 weeks

Vaginal discharge (75% of patients)

Intermenstrual or heavier than usual menstrual bleeding (40%)

Fever (40%)

Nausea and vomiting are common

Lower abdominal tenderness and guarding

Pelvic examination reveals mucopurulent cervical discharge, exquisite tenderness on movement of the cervix, and thickened adnexae

A pelvic mass may be palpable

Patients with tubo-ovarian abscess are more apt to have high fever and prominent leukocytosis; abdominal pain and tenderness are found in over 90% of these patients

Diagnosis

Often there is a history of a new or recent sexual partner

Typical clinical picture

Laparoscopy (mixed aerobes and anaerobes can be recovered from infected Fallopian tubes or pelvic peritoneum in at least one-third of women with acute PID). Gram stain and aerobic and anaerobic culture should be performed on material that is recovered. Optimum anaerobic transport is important.

Blood cultures

Ultrasound, CT scan are very useful

Culdocentesis with Gram stain and appropriate cultures

Therapy

The need for early surgical intervention in the treatment of tubo-ovarian abscess is controversial. Failure of response to an appropriate antimicrobial regimen, however, is an indication that surgery will be required.

Outpatient antimicrobial regimens that are suitable include amoxicillin (3 g PO) plus probenecid as a single dose, followed by doxycycline for 10-14 days, or trimethoprim/ sulfamethoxazole plus clindamycin, or a single 2 g dose of cefoxitin plus probenecid followed by doxycycline for 10-14 days, or ampicillin plus metronidazole (for up to 6-8 weeks in the case of tubo-ovarian abscess).

Inpatient antimicrobial regimens include the following choices: metronidazole plus doxycycline; clindamycin plus gentamicin; cefoxitin plus doxycycline; ceftizoxime plus doxycycline; or ampicillin/sulbactam plus doxycycline.

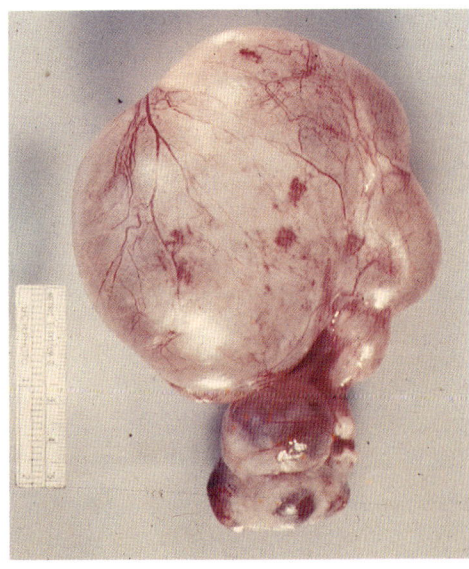

Figure 51.
Large tubo-ovarian abscess.

Septic Abortion

Etiology

Any of the lower genital tract organisms may be involved in postabortion endometritis, but anaerobes and group B streptococci are commonly isolated. Anaerobes of special concern are *B. fragilis* and *C. perfringens*. Other nonanaerobes of concern are *N. gonorrhoeae* and *E. coli*.

Underlying Problems and Complications

During illegal abortions, unsterile instruments are often passed into the uterus several times, resulting in seeding with a variety of organisms.

Even with an abortion carried out under optimum conditions endogenous flora from the vagina and cervix may be introduced into the uterus.

Complications include spread of infection from the uterine cavity to the parametrial tissues and adnexae and perforation of the uterus which allows spread of the infection into the peritoneal cavity. Septicemia is a dreaded complication.

Clinical Features

Fever

Severe uterine cramps

Vaginal bleeding is common and clots may be passed

There may be foul vaginal discharge

Pelvic examination may reveal a tender uterus and, at times, placental fragments, fetal parts, foreign objects, or purulent drainage from a patulous cervical os. Pelvic examination may also reveal that the infection has spread beyond the confines of the uterus.

With complicating sepsis, there may be shaking chills, hemoglobinemia, hemoglobinuria, hypotension, and anuria

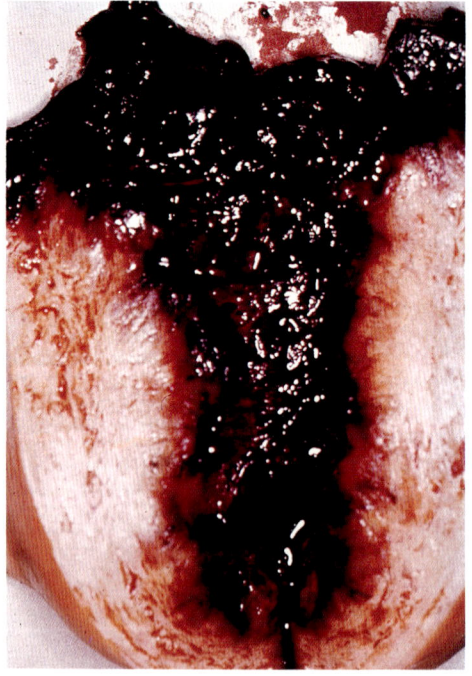

Figure 52.
Autopsy specimen of uterus following criminal abortion. Note hemorrhagic endometritis and gas in myometrium.

Diagnosis

Women frequently deny having a criminal abortion

Typical clinical picture

Gram stain, aerobic and anaerobic culture of material from the uterine cavity, preferably obtained through a sheathed device such as the Pipelle®

Blood cultures (positive in 40 to 60% of patients, primarily with obligate anaerobes recovered)

Roentgenographic examination of the abdomen looking for intra- or extrauterine foreign bodies and gas. Layering of gas in the myometrium gives a characteristic onion-skin appearance and is indicative of extensive necrosis of the uterus

Icteric serum and mahogany-colored urine are poor prognostic signs

Therapy

Initial treatment should be directed toward the accompanying shock, DIC, and intravascular hemolysis, if these are present.

The key to treatment is early evacuation of the uterine cavity. With layering of gas in the myometrium, laparotomy with hysterectomy and, perhaps, excision of other affected organs is indicated.

Antimicrobial therapy should be intensive. Among the regimens that are suitable are ampicillin/sulbactam with or without gentamicin, metronidazole plus ampicillin with or without gentamicin, ticarcillin/clavulanate, imipenem, and chloramphenicol.

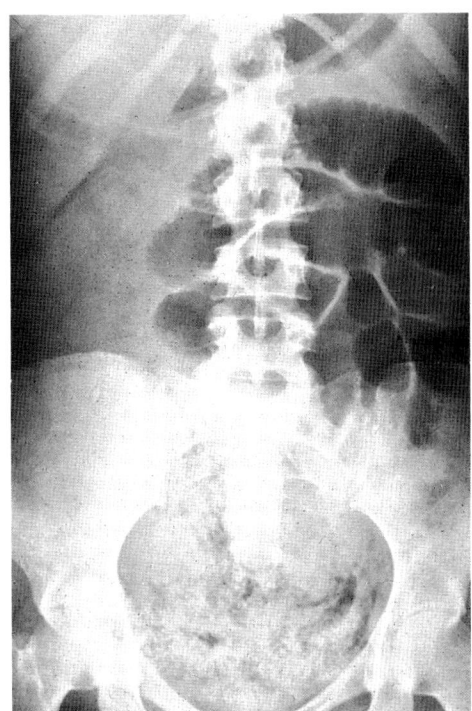

Figure 53.
Flat plate of abdomen showing layers of gas in myometrium (enlarged uterus).

Bite-Wound Infection

Etiology

Anaerobes are found in 30 to 40% of dog-bite wounds (an average of 2.1 anaerobes/ wound compared with an average of 4.1 nonanaerobes). The most common anaerobes encountered in animal bite infections are *Bacteroides* spp.; *Fusobacterium* spp. and *Peptostreptococcus* spp. are also found regularly.

Pasteurella multocida is an important pathogen in infections following dog and cat bites. Another nonanaerobe that seems to be unique to dog-bite wounds is *S. aureus* (found in 10 to 25% of such wounds). It is likely that some of these isolates actually represent *Staphylococcus intermedius*.

Clostridia, including *C. perfringens*, are found in snake-bite wounds and, perhaps, in wounds related to shark bites.

Anaerobes are found in 50 to 60% of infected human bite (or clenched fist injury) wounds: *Bacteroides*, *Fusobacterium*, and *Peptostreptococcus* spp. are the most prevalent, but representatives of several other anaerobic genera also have been found.

Eikenella corrodens is an important pathogen in human bite wounds.

Clinical Features

There is often significant tissue damage with both animal and human bites; jagged lacerations and significant tissue necrosis may be found deep in the wound.

In the case of clenched-fist injuries there may be damage to tendons and the joint capsule; this makes the prognosis poorer.

Wound discharge is typically purulent and not uncommonly foul smelling.

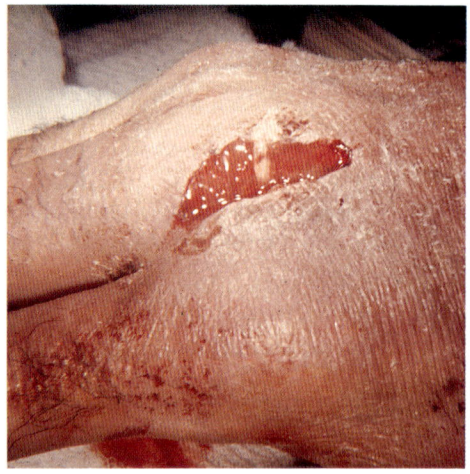

Figure 54.
Human "bite wound" (clenched fist injury) of hand. Note jagged laceration and severed tendon in wound.

Diagnosis

Typical clinical picture

History of bite or fist injury related to a fight

Gram stain and aerobic and anaerobic culture should be obtained; optimum anaerobic transport is essential

Aerobic and anaerobic blood cultures should be done, but are usually negative

Radiographic and other studies may be indicated to rule out involvement of underlying bone (osteomyelitis) and to detect gas in the soft tissues

Therapy

Surgical treatment is important. There should be thorough debridement, drainage and cleansing of the wound. Repair to deeper tissues, such as tendons and joint capsules, should be undertaken.

Amoxicillin/clavulanic acid and ampicillin/sulbactam represent excellent choices for bite wound infections of all types. Metronidazole plus ampicillin represents another good choice.

Penicillin G is the drug of choice for infections with *P. multocida*. Isolates of this organism are resistant to erythromycin, vancomycin, and clindamycin; dicloxacillin, cephalexin, and cefaclor probably would not achieve sufficient levels to treat *P. multocida* infection reliably.

In the case of dog bites, therapy directed against *S. aureus* may be indicated.

Depending on the nature and circumstances of the bite, rabies and tetanus prophylaxis should be considered.

Skin and Soft Tissue Infections

Anaerobic infections of the skin and soft tissue may develop in areas whose integrity has been interrupted by surgical wounds, injuries, bites and various pathological processes, and in areas of ischemia. While the majority of these infections are not life threatening, some of them are and these must be recognized as early as possible.

Clues to Anaerobic Infections of Skin and Subcutaneous Tissue

Gas in tissues or discharges

Abscess formation

Tissue necrosis, gangrene

Foul odor

Sinus tract formation

Sulfur granules (actinomycosis)

Impaired blood supply to site

History of bite

History of "skin popping"

Blocked apocrine glands

Unique morphology on Gram stain of exudate

No growth on routine culture

Severe, deep-seated soft tissue infections are all characterized by marked toxemia and pain

Classification

TYPES OF ANAEROBIC INFECTIONS INVOLVING PRIMARILY SKIN AND SKIN STRUCTURES

Cellulitis (including necrotizing cellulitis)

Infected cutaneous ulcers

Subungual infection

Paronychia

Infected sebaceous or inclusion cysts

Pyoderma

Hidradenitis suppurativa

Subareolar breast abscess

Cutaneous abscess

Cutaneous actinomycosis (rare)

Tropical ulcer

Perioral dermatitis

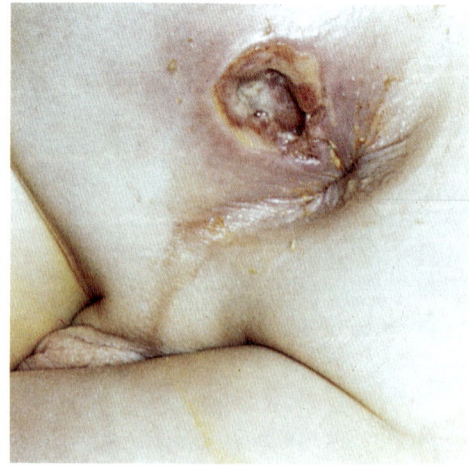

Figure 55.
Perirectal abscess.

62

TYPES OF ANAEROBIC INFECTIONS OF SUBCUTANEOUS TISSUE +/- SKIN

Subcutaneous abscesses (including those due to skin popping)

Anaerobic cellulitis, "gas abscess"

Infected diabetic and other foot ulcers

Perirectal abscess

Infected pilonidal cyst/sinus

Infected decubitus ulcers

Surgical wound infections

Burn wound infections

Bite infections (human and animal)

Chronic undermining ulcer of Meleney

Bacterial synergistic gangrene

Infected malignant ulcers

Noma (cancrum oris)

TYPES OF ANAEROBIC INFECTIONS OF DEEPER TISSUES

Necrotizing fasciitis

Clostridial myonecrosis (nontraumatic and other)

Nonclostridial anaerobic myonecrosis (synergistic necrotizing cellulitis)

Anaerobic streptococcal myositis

Infected vascular gangrene

Muscle abscess

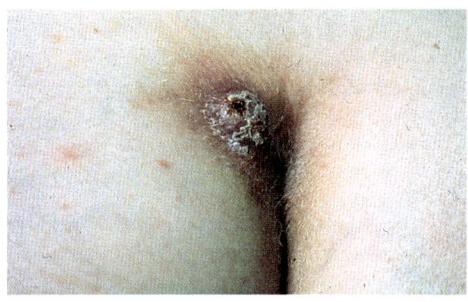

Figure 56.
Pilonidal abscess.

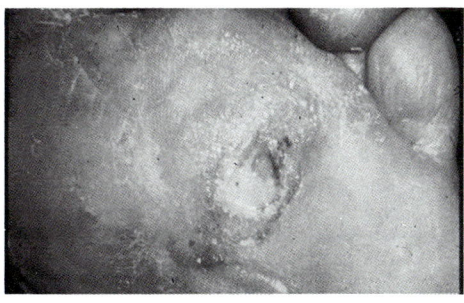

Figure 57.
Purulent exudate in the base of a trophic ulcer of the ball of the foot in a diabetic with neuropathy.

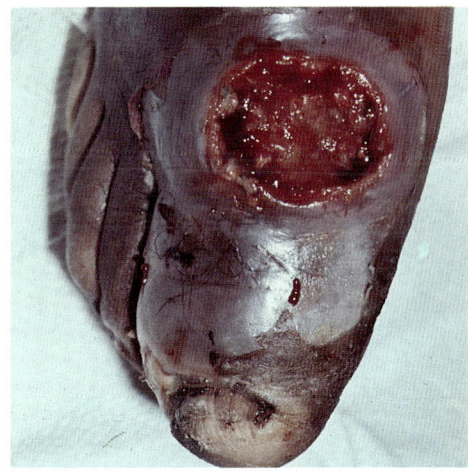

Figure 58.
Infected foot ulcer in a diabetic with peripheral vascular disease and neuropathy.

Hidradenitis Suppurativa

Chronic suppurative disease of apocrine glands

Scarring, sinus tract formation common

Axillae, genital, and perineal areas most commonly involved

Infection involves anaerobes (especially *Bacteroides*, *Peptostreptococcus*), *S. aureus*, streptococci, and sometimes nonanaerobic gram-negative bacilli

Therapy difficult. Antimicrobials, moist heat, surgical drainage, and in severe, chronic cases, surgical excision, and skin grafting

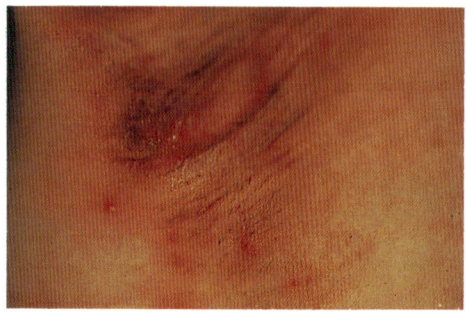

Figure 59.
Hidradenitis suppurativa, chronic, of axilla. Small superficial abscess present.

Anaerobic Cellulitis

Gradual onset
Little change in skin
Little or no exudate
Abundant gas
Foul odor
No change in muscle
Infection involves *B. fragilis*, other *Bacteroides*, *Clostridium*, *Peptostreptococcus*, coliforms, streptococci
Treatment: debridement, drainage, and antimicrobials

Diabetic Foot Infections

Table 5. BACTERIOLOGY OF 32 DEEP-TISSUE SPECIMENS, INFECTED DIABETIC FEET

	Aerobes	Anaerobes	Aerobes + Anaerobes
No. Isolates/specimen	2.8	2.0	4.8
Count (log/g)	6.0*	7.5*	6.6

ANAEROBES RECOVERED

Bacteroides fragilis group	10	*Peptostreptococcus* spp.	9
Bacteroides spp.	12	*Clostridium* spp.	11
Fusobacterium spp.	2	Non-spore-forming gram-positive rods	5
Unidentified gram-negative rod	1		

From Sapico FL. Foot infections. In SM Finegold, WL George (eds.), *Anaerobic Infections in Humans.* Academic Press, San Diego, 1989.
* P < 0.05, Student's t test

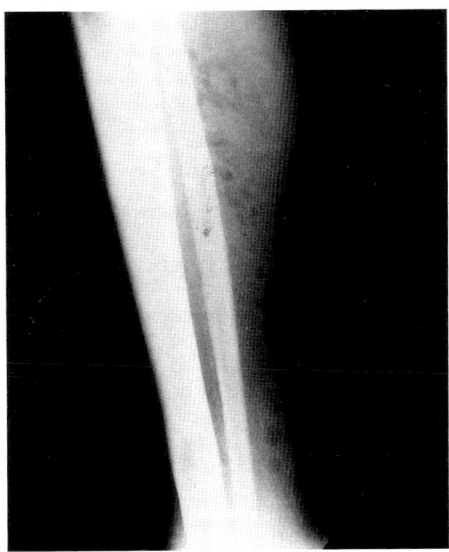

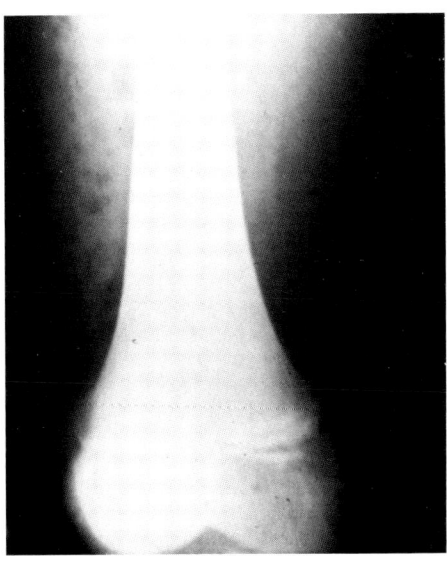

Figure 60.
Marked accumulation of gas and soft tissue swelling of lower leg in diabetic with anaerobic cellulitis.

Figure 61.
Anaerobic cellulitis. Gas-forming infection in thigh of young girl; seeded during sepsis with Fusobacterium necrophorum (original process was Vincent's angina).

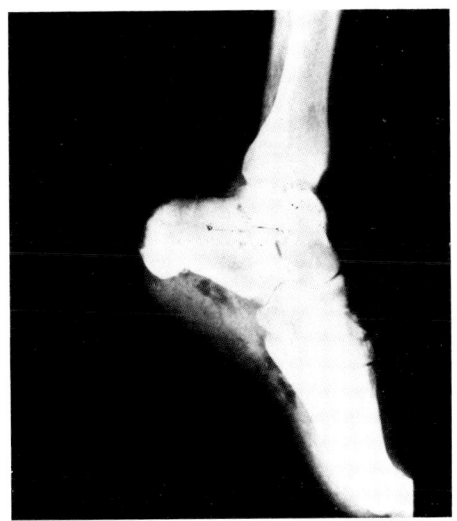

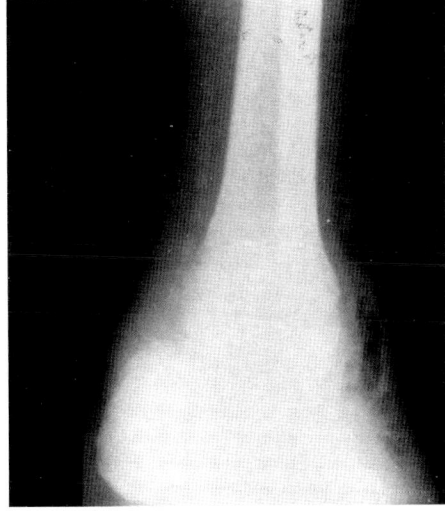

Figure 62.
Anaerobic cellulitis, arch of foot. Note gas in soft tissues. Patient was diabetic.

Figure 63.
Anaerobic cellulitis, ankle with much gas present. Note calcified vessel parallel to tibia.

Infected Decubitus Ulcer

Basic problem is pressure sore

Occur primarily over sacrum, hips, and other bony prominences

Necrotic tissue becomes nidus of infection

Deeper tissues, including fascia, muscle, tendon, and bone may become involved

Infection tends to undermine so that drainage becomes a problem

Bacteremia, especially with *B. fragilis* and other anaerobes, is a relatively frequent complication

Infecting flora includes all types of anaerobes (with *B. fragilis* group usually dominant), *Enterobacteriaceae*, *Pseudomonas*, *S. aureus*, and enterococci and other streptococci

Treatment includes elimination of pressure, debridement and drainage, and antimicrobials. Suitable drugs depend on infecting organisms but include ampicillin/sulbactam, ticarcillin/clavulanate, cefoxitin, and others. Topical metronidazole is effective against the anaerobic flora

Meleney's (Chronic Undermining) Ulcer

Slowly progressive infection of subcutaneous tissue with ulceration of skin leaving undermined bridges

Periphery erythematous and tender

Etiology: microaerophilic (occasionally anaerobic) hemolytic streptococcus

Treatment: debridement and drainage, antimicrobials

Progressive Bacterial Synergistic Gangrene

Usually postoperative complication, especially following abdominal or thoracic surgery

Lesion very painful and tender; extends slowly

Typical "triple zone" lesion — central necrotic zone, middle zone of purplish discoloration, outer zone of less intense erythema

Etiology: microaerophilic (or anaerobic) *Streptococcus* plus *S. aureus* (or gram-negative rod)

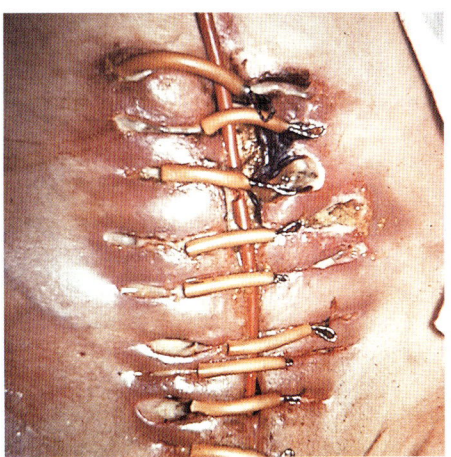 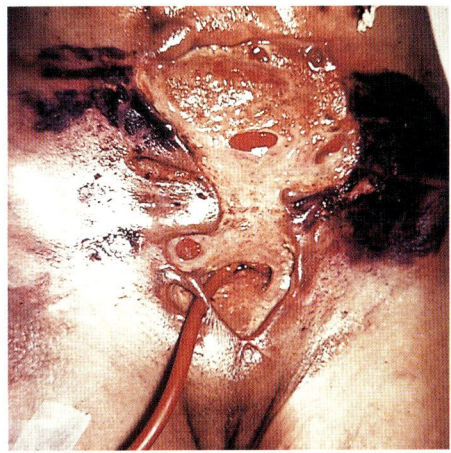

Figure 64.
Progressive bacterial synergistic gangrene in abdominal surgical wound with retention sutures. Lesion shows typical "triple zone."

Figure 65.
Same patient as in Fig. 64. Wound dehiscence with ensuing evisceration.

Severe Anaerobic Soft Tissue Infections

(ALL CHARACTERIZED BY MARKED TOXEMIA AND PAIN)

Necrotizing Fasciitis

Acute onset

Erythema, cellulitis, ecchymosis, extensive undermining of skin and subcutaneous tissue, gangrene; anesthesia late

Serosanguinous exudate

Gas uncommon

Foul odor

Muscle intact; fascia necrotic

Infection involves *B. fragilis*, other *Bacteroides*, *Clostridium*, *Peptostreptococcus*, *S. pyogenes*, *S. aureus*

Treatment: immediate surgical intervention, antimicrobials

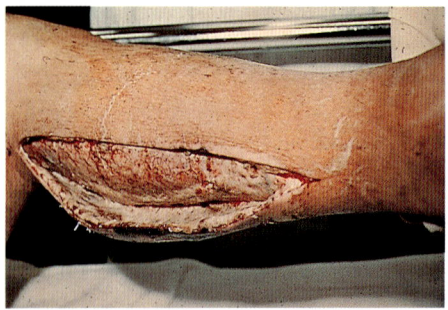

Figure 66.
Necrotizing fasciitis, calf, following injury. Note necrotic fascia with purulence and large potential space between fascia and subcutaneous tissue.

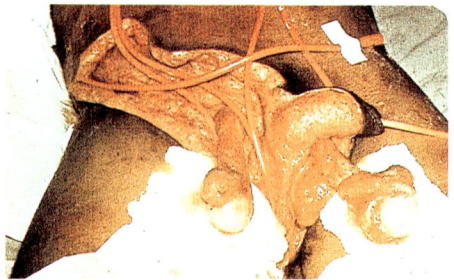

Figure 67.
Fournier's gangrene. Necrotizing fasciitis that began on the scrotum of this young man and spread quickly to involve penile shaft, lower abdominal wall, upper thigh, and back. Wound has been thoroughly debrided and good granulation tissue is present. Soon after this picture was taken, the denuded testes were implanted in the thighs, and a skin graft was applied successfully.

Clostridial Myonecrosis

Acute onset

Skin tense, often pale or bronzed, bullae may be present

Exudate often profuse, serosanguinous

Gas not pronounced except terminally

Slight, sweetish odor

Marked change in muscle

Infection involves *C. perfringens*, *C. histolyticum*, *C. septicum*, *C. novyi*, *C. bifermentans*

Treatment: early, extensive surgical debridement of *all* involved tissue. Antimicrobial therapy. Consider hyperbaric oxygen.

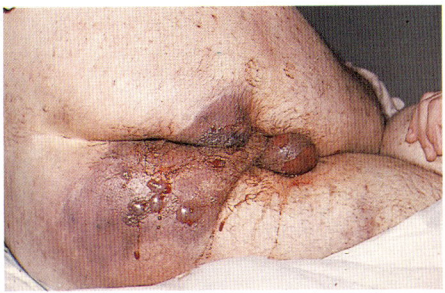

Figure 68.
Gas gangrene (clostridial myonecrosis)
of buttock. Note hemorrhagic discolor-
ation, hemorrhagic bullae, and thin
hemorrhagic fluid.

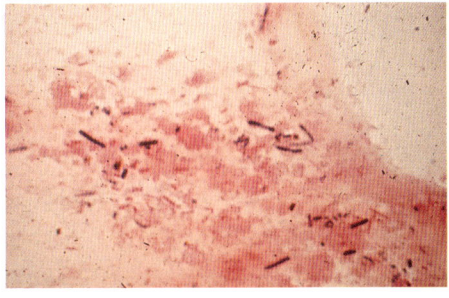

Figure 69.
Gram stain from the patient in Fig. 68.
Note large, gram-positive rods
(C. perfringens), small, gram-negative
rods (E. coli) and destruction of
leukocytes (by clostridial toxin).

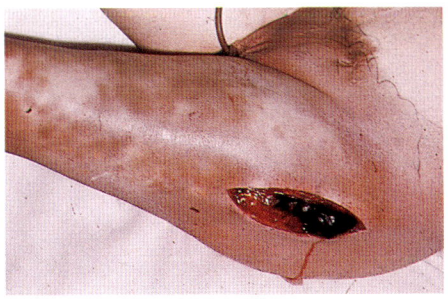

Figure 70.
Clostridial myonecrosis following
hip pinning. Note marked bronzed
discoloration (compare with other leg)
and extension to lower abdominal wall
(marked with ink).

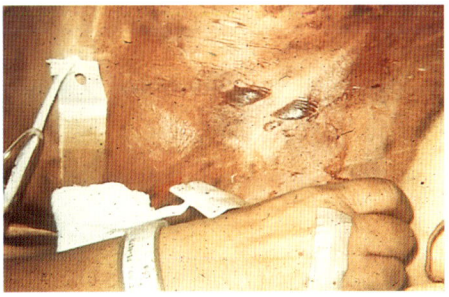

Figure 71.
Clostridial myonecrosis, abdominal
wall following abdominal surgery. Note
bronzed discoloration of skin and
hemorrhagic bullae.

Synergistic Nonclostridial Anaerobic Myonecrosis

Acute onset

75% of patients have diabetes mellitus

Skin has scattered blue-gray necrosis with normal intervening areas

"Dishwater pus" type of exudate

Gas not pronounced; present in 25% of patients

Foul odor

Marked change in muscle

Infection involves *Bacteroides*, *Peptostreptococcus*, aerobic or facultative gram-negative rods

Treatment: As for clostridial myonecrosis

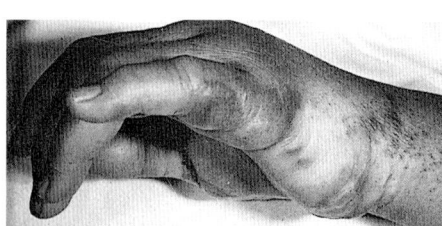

Figure 72.
Synergistic nonclostridial anaerobic myonecrosis. Patchy, blue-gray discoloration of lateral aspect of hand. Underlying necrosis is much more extensive than it appears on the surface.

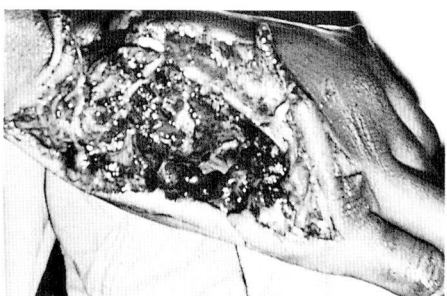

Figure 73.
Same patient as in Fig. 72, after wound exploration.

Anaerobic Streptococcal Myositis

Subacute or insidious onset

Skin tense, often coppery tinge

Profuse, seropurulent exudate

Gas not extensive

Slight, often sour odor

Muscle pale and soft initially, later purple and friable

Infection involves *Peptostreptococcus* plus *S. pyogenes* and/or *S. aureus*

Treatment: debridement and drainage, antimicrobials

Infected Vascular Gangrene

Gradual onset

Skin appearance varies with extent of vascular disease

Exudate nil

Abundant gas

Foul odor

Dead muscle (vascular gangrene)

"Saprophytic" invasion of dead tissue. Various anaerobes and nonanaerobes are involved.

Treatment: amputation or debridement

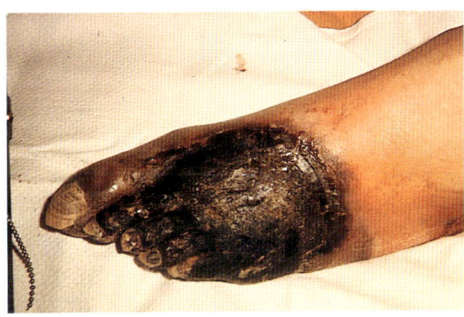

Figure 74.
Infected vascular gangrene.

Table 6. DIFFERENTIATING CHARACTERISTICS OF CERTAIN SIMILAR ANAEROBIC INFECTIONS OF SKIN AND SOFT TISSUE[a]

	Clostridial Myonecrosis (Gas Gangrene)	Synergistic Non-Clostridial Anaerobic Myonecrosis	Anaerobic Streptococcal Myositis	Infected Vascular Gangrene	Necrotizing Fasciitis	Anaerobic Cellulitis
Incubation	Usually less than 3 days	Variable. 3–14 days	3–4 days	More than 5 days. Usually longer	1–4 days	Almost always more than 3 days
Onset	Acute	Acute	Subacute or insidious	Gradual	Acute	Gradual
Toxemia	Very severe	Marked	Severe only after some time	Nil or minimal	Moderate to marked	Nil or slight
Pain	Severe	Severe	Variable, as a rule fairly severe	Variable	Moderate to severe	Absent
Swelling	Marked	Moderate	Marked	Often marked	Marked	Nil or slight
Skin	Tense, often very pale	Minimal change	Tense, often with a coppery tinge	Discolored, often black & desiccated	Pale red cellulitis	Little change
Exudate	Variable, may be profuse: serous & blood stained	"Dishwater" pus	Very profuse, seropurulent	Nil	Serosanguinous	Nil or slight
Gas	Rarely pronounced except terminally	Not pronounced: present in 25% of cases	Very slight	Abundant	Usually not present	Abundant
Smell	Variable, may be slight, often sweetish	Foul	Very slight, often sour	Foul	Foul	Foul
Muscle	Marked change	Marked change	At first little change except edema	Dead	Viable	No change

[a] Facultative bacteria may also produce some of these infections.

Adapted from MacLennan JD. The histotoxic clostridial infections of man. *Bacteriol. Rev.* 26:177–276, 1962.

Therapy of Skin and Soft Tissue Infections

INFECTIONS REQUIRING EXTENSIVE SURGICAL INCISION AND DRAINAGE

Necrotizing fasciitis

Anaerobic streptococcal myositis

Crepitant cellulitis (clostridial and nonclostridial)

INFECTIONS REQUIRING EXCISION OF TISSUE

Progressive bacterial synergistic gangrene

Chronic burrowing ulcer

Clostridial myonecrosis

Synergistic necrotizing cellulitis

ANTIMICROBIAL THERAPY OF SKIN AND SOFT TISSUE INFECTIONS

Important adjunct to surgical management

Choice of agents depends on specific bacteriology

The most potent, most consistently active drugs should be used in serious, deep-seated infections. Included in this category are metronidazole, ticarcillin/clavulanate, ampicillin/sulbactam, imipenem, and chloramphenicol. Penicillin G is still active against most clostridia. A significant number of clostridia are resistant to cefoxitin and clindamycin.

Osteomyelitis

Etiology

As many as 40% of cases of osteomyelitis yield anaerobes.

Most cases of osteomyelitis involving anaerobes are mixed with nonanaerobes.

Anaerobes — *Peptostreptococcus*, *Actinomyces*, *Propionibacterium*, *F. necrophorum*, other fusobacteria, *Bacteroides* (including the *B. fragilis* group), and *Clostridium*.

Nonanaerobes include *S. aureus*, streptococci, *Enterobacteriaceae* and *Pseudomonas*.

Underlying Problems

Actinomycosis

Bacteremia (*F. necrophorum*, Vincent's angina)

Peripheral vascular disease, neuropathy, diabetes mellitus

Contiguous focus of infection

Trauma

Prosthetic devices

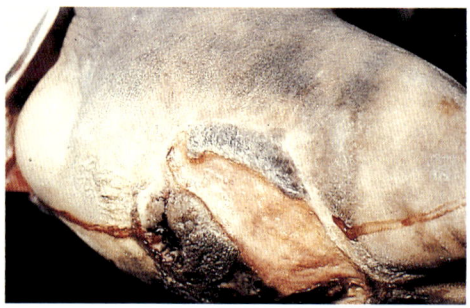

Figure 75.
Diabetes mellitus, Charcot's joint.
Osteomyelitis of bone exposed in base of trophic ulcer.

Clinical Features

One or more of above underlying problems

Often mimics aerobic osteomyelitis

Foul-smelling drainage

Gas in soft tissues

Negative routine cultures

Gram stain of discharge showing morphology typical of certain anaerobic organisms

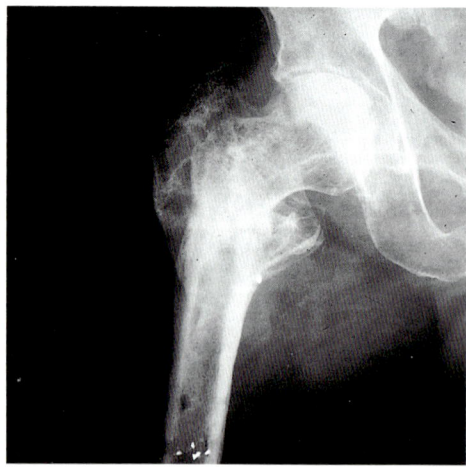

Figure 76.
Radiograph showing osteomyelitis of hip following trauma. Peptostreptococcus *was recovered in pure culture.*

Diagnosis

Clinical features as described above

Radiographs

Bone, gallium scans; indium-labeled white blood cell scan

Bone biopsy with histology and culture (best specimen for culture)

Curettings from deep in sinus tract next best specimen for culture

Culture of pus aspirated by syringe or on swab (latter is less desirable)

Blood culture

CT scan, MRI

Therapy

Debridement, drainage

Removal of foreign bodies

Improved vascularization when feasible

Antimicrobial agent(s) — choice depends on bacteriology. Likely choices include ampicillin/sulbactam, ceftizoxime, cefoxitin, clindamycin, metronidazole, penicillin, ticarcillin/clavulanate, and imipenem.

Antimicrobial therapy must be prolonged

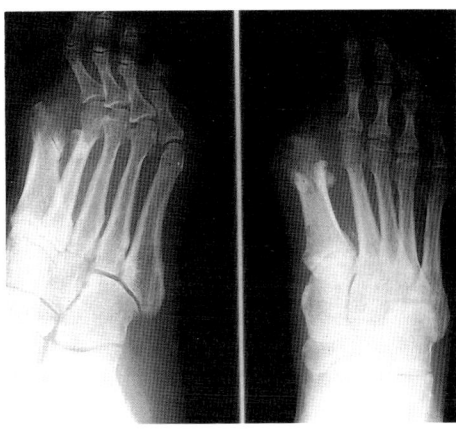

Figure 77.
Radiographs of foot showing osteomyelitis of first metatarsal (toe previously amputated). Note irregular bone edge and periosteal thickening.

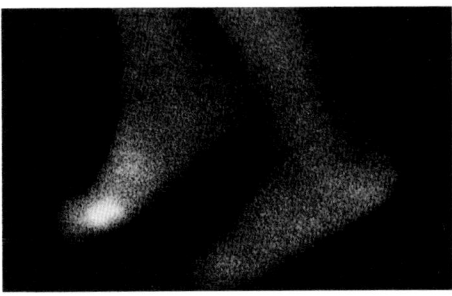

Figure 78.
Gallium isotope scan consistent with osteomyelitis of the great toe.

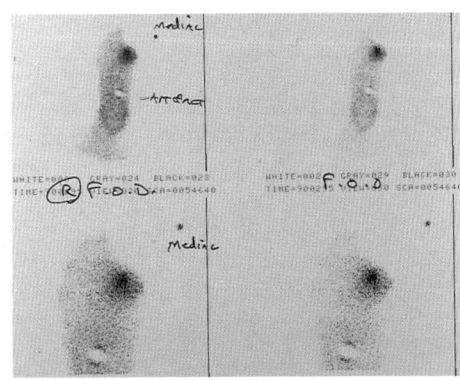

Figure 79.
Bone scan consistent with osteomyelitis of the first metatarsal (same patient as in Fig. 77).

Purulent Arthritis

Etiology

Only 1% of cases of septic arthritis have yielded anaerobes; however, almost 20% have had negative cultures suggesting that some of these were probably due to anaerobes.

F. necrophorum is most common anaerobic isolate, with anaerobic cocci second. Others include *Bacteroides* (*B. fragilis* and others), other fusobacteria, clostridia, and *Actinomyces*.

Underlying Problems

Hematogenous spread

Previous injury, arthritis in joint

Hemoglobinopathies

Trauma

Contiguous osteomyelitis

Prosthetic devices

Steroid injection into joint

Clinical Features

Underlying problems as outlined above

Septic arthritis with negative routine culture

Intra-articular gas

Foul odor to joint fluid

Gram stain of joint fluid showing morphology typical of certain anaerobic organisms

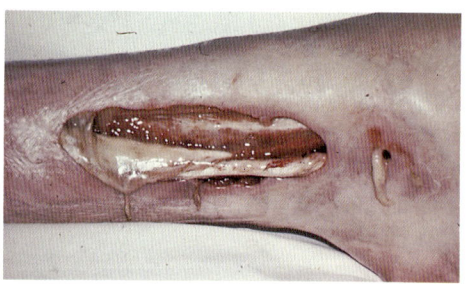

Figure 80.
Anaerobic wound infection following compound fracture. Patient developed purulent arthritis of hip with draining sinus.

Diagnosis

Clinical features as noted above

Culture of joint fluid

Blood culture

Therapy

Drainage (by repeated arthrocentesis)

Joint immobilization early

Antimicrobial therapy, as indicated by joint fluid culture; usually for at least two months

Removal of prosthesis if one is present

Actinomycosis

Etiology

Actinomyces israelii most common infecting organism. The other species of *Actinomyces* as well as *Propionibacterium propionicus* (*Arachnia propionica*), *Eubacterium nodatum, E. timidum*, and *E. brachy* may be involved. Other bacteria are usually involved as well — included are *Actinobacillus actinomycetemcomitans*, *Bacteroides* (especially pigmented species), *Porphyromonas, Haemophilus* spp., *E. corrodens*, various streptococci, and *Enterobacteriaceae*.

Underlying Problems

Tooth extraction

Oral, bowel, or gallbladder surgery

Trauma

Aspiration

Chronic bronchitis

Emphysema

Bronchiectasis

Intrauterine contraceptive device

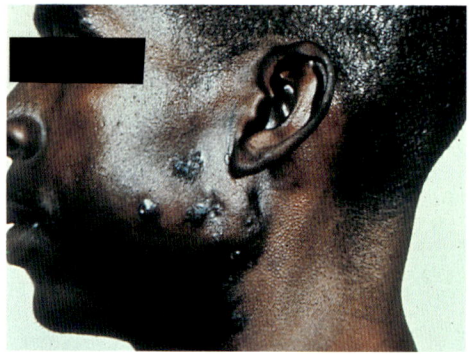

Figure 81.
Cervicofacial actinomycosis. Note swelling of jaw (very firm to touch) and multiple sinuses of the face and neck.

Clinical Features

Cervicofacial disease (half of all cases) manifests as a chronic fluctuant mass of woody consistency in the parotid or submandibular area. Sinus tracts discharging "sulfur granules" are not uncommon.

Thoracic disease — consolidated or cavitary pneumonia, extension to pleural space and chest wall (including ribs) to produce empyema and/or chest wall mass or draining sinus or fistula. May extend into mediastinum to involve great vessels, pericardium, myocardium, or endocardium.

Abdominal disease — intra-abdominal mass. May spread to diaphragm, liver, kidney, various other organs, retroperitoneum, perirectal tissues. Portal bacteremia may occur.

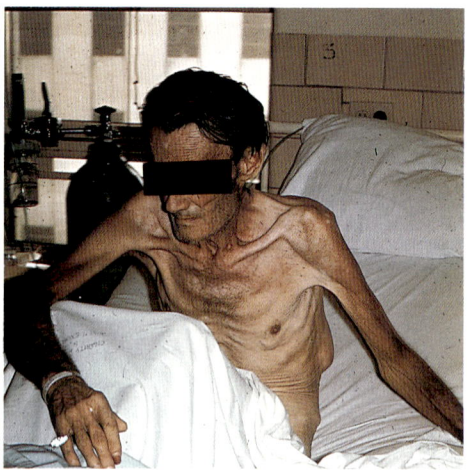

Figure 82.
Patient with pulmonary and pleural actinomycosis with soft tissue abscess of chest wall.

Infected IUDs

CNS disease is rare; included are brain abscess, subdural empyema, epidural abscess, and osteomyelitis of skull

Lacrimal canaliculitis, sinusitis, and involvement of palate, salivary glands, larynx, trachea, and hypopharynx occur uncommonly

Diagnosis

Characteristic clinical features

Sulfur granules (strain pus through cheesecloth to find these)

Gram stain and culture of sulfur granules, pus. Branching thin gram-positive rods or filaments are characteristic (rule out *Nocardia*, which has a similar appearance but is weakly acid fast).

Biopsy with characteristic histology

Direct fluorescent antibody test is available from CDC, but doesn't cover all species of *Actinomyces*

Radiographs

CT, MRI scans; radionuclide scans

Therapy

Penicillin G in high dosage (18 million units/day) is drug of choice. Prolonged treatment (at least several months) is essential. Tetracycline is an effective alternative, as is clindamycin (except that *Actinobacillus* is resistant to the latter). Metronidazole has poor activity against *Actinomyces*.

Surgery is extremely important; debridement, drainage are essential and excision of sinus tracts may be required.

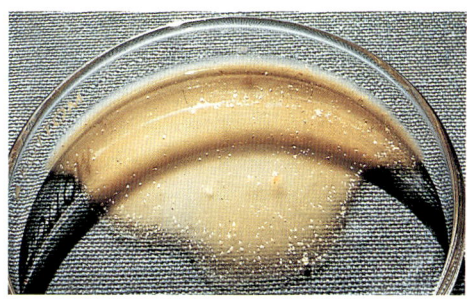

Figure 83.
Pus with "sulfur granules" aspirated from chest mass of patient in Fig. 82.

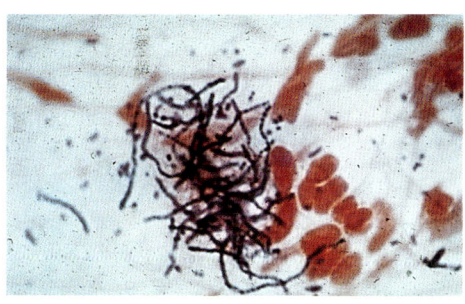

Figure 84.
Gram stain showing characteristic morphology of Actinomyces.

Subareolar Breast Abscess

Etiology

In contrast to puerperal breast abscess, which often involves *S. aureus*, subareolar breast abscess (nonpuerperal) is primarily an anaerobic infection. The anaerobes commonly involved are *Bacteroides* (pigmenters, *B. oralis* group, *B. bivius*, *B. disiens*, *B. ureolyticus* group), *Porphyromonas*, *Fusobacterium*, and anaerobic cocci (especially *Peptostreptococcus*).

Underlying Problems

Inverted or retracted nipples

Plugging of ducts lined with squamous epithelium by keratin

Gynecologic manipulation

Clinical Features

Chronic, recurrent (over many years)

Draining sinuses may occur

Foul-smelling discharge

Subareolar in location

Diagnosis

Classical clinical picture

Aerobic and anaerobic culture of pus or sinus drainage

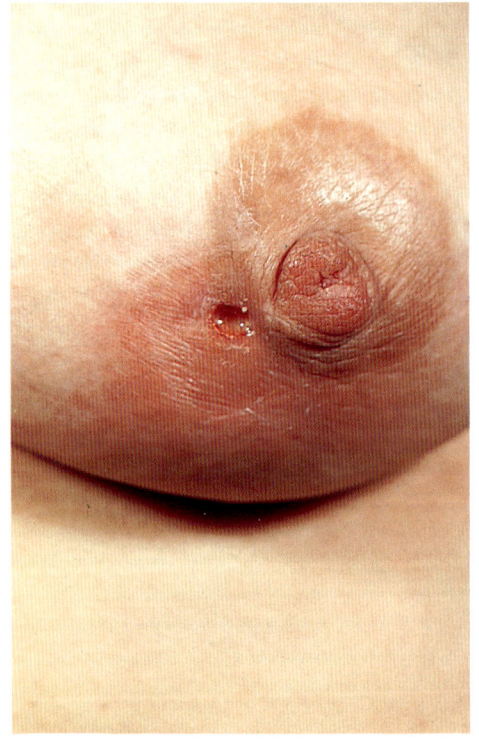

Figure 85.
Subareolar breast abscess with
fluctuant abscess beneath nipple and
chronic sinus draining foul-smelling pus.

Therapy

Primary therapy should be surgical; this should include duct excision, not just drainage of a fluctuant area

Antibiotics are adjunctive. Oral agents should be suitable — metronidazole, amoxicillin/clavulanate, or clindamycin

Tetanus

Etiology

Clostridium tetani

Underlying Problems

Complication of lacerations, open fractures, burns, frostbite, abrasions, puncture wounds, hypodermic injections, childbirth (infection of the umbilical stump in the newborn), and surgery on the gastrointestinal tract

Clinical Features

Incubation period usually <14 days from time of injury; up to 54 days

Usually generalized; occasionally presents as a localized process

Trismus

Risus sardonicus

Spasm of pharyngeal muscles

Stiff neck, opisthotonos

Tightness of chest muscles; rigidity of abdominal wall, back and limbs

Generalized tonic convulsions — excited by any sudden jar or sound

Spasm of laryngeal and respiratory muscles — danger of asphyxia

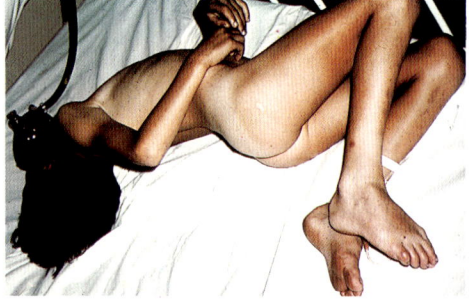

Figure 86.
Tetanus with opisthotonos.

Diagnosis

Distinctive clinical picture

Laboratory tests of little help

Therapy

Antitoxin — tetanus immune globulin (TIG), 500 to 6000 U deep IM in proximal portion of wounded extremity or in gluteal muscles

Intrathecal antitoxin not proven to be of value

Constant, expert, vigilant nursing care

Muscle relaxants — diazepam

Analgesics, sedatives

Surgical wound care

Antibiotics as indicated for complications

Tracheostomy when indicated

Botulism

Etiology

Clostridium botulinum, toxin types A, B, E, F. In infant botulism, a *C. barati* strain producing type F botulinal neurotoxin and a *C. butyricum* strain producing type E toxin have been isolated.

Underlying Problems

Ingestion of preformed toxin in food contaminated with botulinal spores that subsequently germinate, under appropriate conditions (particularly home canning of food, smoked fish), leading to toxin production. This is typical mechanism of botulism in adults.

In wound botulism, a wound is colonized or infected with *C. botulinum* which produces toxin locally that is then absorbed.

In infant botulism, spores of the organism are ingested; because of the incompletely developed intestinal flora of the young infant, the *C. botulinum* spores can colonize, germinate and produce toxin in the gut, from where it is absorbed.

Spores have been found in honey and corn syrup.

Clinical Features

Bilateral cranial nerve impairment of acute onset

Dysphagia, dry mouth, diplopia, dysarthria, blurred vision

Symmetrical descending weakness or paralysis subsequently; respiratory arrest may occur

Nausea, vomiting may be seen

Ileus, constipation

Incubation period usually <3 days; may be up to 8 days or longer

Infected wound in the case of wound botulism

In infant botulism, constipation, weakness and lethargy, feeding difficulty, weak cry, ptosis, fatigability with repetitive motor activity

Autonomic nervous system dysfunction

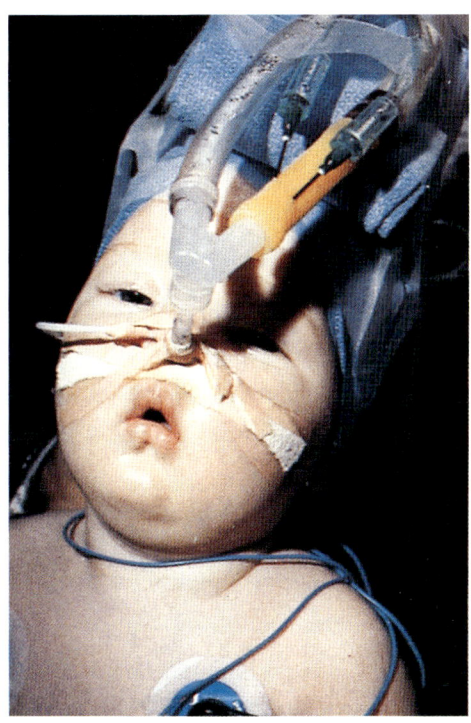

Figure 87.
Infant botulism. Patient has ptosis, dilated pupils. Nasogastric feeding tube is in place, and patient is on respirator (not shown).

Diagnosis

Typical clinical picture, epidemiology

Identification of botulinal neurotoxin in serum, stool or implicated foods, or recovery of *C. botulinum* from stool or wound

Normal cerebrospinal fluid

Characteristic electromyographic findings — brief, small abundant potentials with facilitation on repetitive stimulation

Therapy

Trivalent (ABE) botulinal antitoxin for adult botulism only — give as soon as possible

Penicillin for wound botulism only

Supportive care extremely important — especially ventilatory support

Clostridium perfringens Food Poisoning

Etiology

Enterotoxin-producing Type A *C. perfringens*

Underlying Problems

Food-borne, primarily related to meat or meat products (including poultry)

Meat cooked, allowed to cool, then inadequately reheated prior to ingestion

Primarily associated with restaurants or institutions; seldom home-associated

Most common in spring and fall

Clinical Features

Incubation period 8-12 hours (extremes of 6-24 hours)

Mild, self-limited enteritis (duration 24 hours)

Diarrhea, crampy abdominal pain

Stools liquid; no blood or mucus

Nausea in 25% of patients, vomiting in 9%

Attack rate high ($>$50%)

Diagnosis

Typical clinical and epidemiologic features

Demonstration of $>10^5$ C. perfringens/g of implicated food

Same serotype of C. perfringens in patients' stools, in counts $>10^6$/g

Immunologic detection of enterotoxin in feces

Therapy

Generally none is needed

Fluid replacement may be required

Antimicrobial-Associated Pseudomembranous Colitis

Etiology

Clostridium difficile is primary pathogen

Other pathogens include *S. aureus*, *C. perfringens*, and rarely other clostridia

Diarrhea without colitis may be caused by enterotoxigenic *C. perfringens*

Underlying Problems

Antimicrobial therapy — virtually all antimicrobial agents, but most commonly ampicillin, clindamycin, and cephalosporins. Exceptions are parenteral vancomycin and aminoglycosides.

Cancer chemotherapy — methotrexate, 5-fluorouracil, cyclophosphamide, doxorubicin

Nosocomial spread — the spores of *C. difficile* persist in the environment for long periods

Surgery, hypotension

Clinical Features

May follow discontinuation of offending agent — by 3 weeks or longer

Diarrhea (may be bloody or profuse); nausea and vomiting rare

Abdominal cramps, distention common

Fever, leukocytosis

Protein-losing enteropathy

Complications (generally rare) include perforation of colon, toxic megacolon, hypotension, chronic intractable colitis, and death

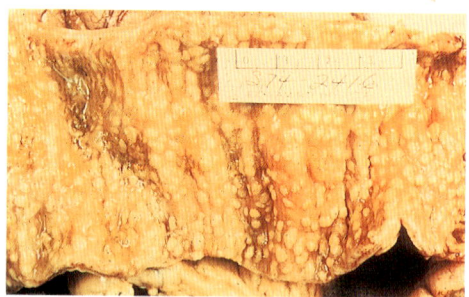

Figure 88.
Antimicrobial agent-induced pseudomembranous colitis. Autopsy specimen showing mucosa studded with classical yellow, elevated plaques.

Diagnosis

Presence of colonic pseudomembrane or pathognomonic yellow elevated plaques definitive diagnostically. May be seen via sigmoidoscope or colonoscope

Demonstrating one of the toxins or antigens of *C. difficile* in the stool is the most common diagnostic test performed. The two tests most commonly employed are a latex agglutination procedure and a cytotoxicity test employing tissue culture cell lines. It should be noted, however, that these are not actually diagnostic, as even high titers of toxin (and high counts of the organism) may be seen in some patients without any symptoms at all who are receiving antimicrobial therapy. Correlation with the clinical picture is therefore important

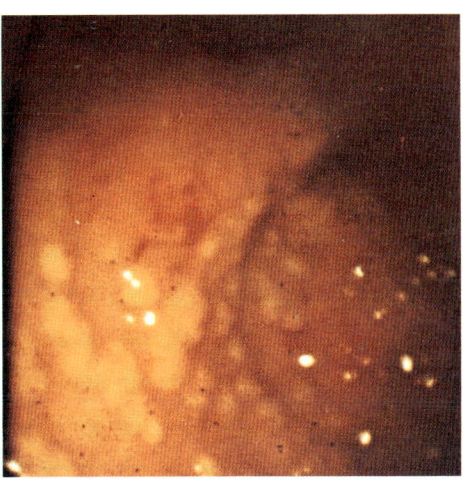

Figure 89.
Colonoscopic view of patient with pseudomembranous colitis revealing the pathognomonic plaques.

Therapy

Simply stopping the offending agent, when feasible, is often adequate to control the disease.

Oral metronidazole, bacitracin, and vancomycin have all been used effectively; vancomycin is very expensive, however. When the patient is unable to take oral medication, intravenous metronidazole is the best antimicrobial regimen.

The relapse rate is high (20 to 25%), and relapse may not occur for several weeks.

Antimotility agents and corticosteroids are hazardous in this disease and should not be used.

Bile acid-binding resins have not been established as effective.

Bacterial Vaginosis (Nonspecific Vaginitis)

Etiology

Some disagreement among investigators. Anaerobes clearly important, especially *Bacteroides* and *Peptostreptococcus* species. The role of *Mobiluncus* spp. and *Gardnerella vaginalis* is uncertain.

Underlying Problems

Poorly understood

Clinical Features

Gray, white, homogeneous, malodorous vaginal discharge with small bubbles

Little or no discomfort

No inflammation

Diagnosis

Presence of "clue cells" on Gram stain of vaginal discharge — epithelial cells covered with masses of gram-variable rods (*Gardnerella* and *Mobiluncus*). Absence of lactobacilli is characteristic. Few or no polymorphonuclear leukocytes seen

pH of discharge > 4.5

Addition of KOH to discharge results in distinct rotten-fish odor (amines)

Therapy

Metronidazole is drug of choice (500 mg to 1 g every 12 hours × 7 days)

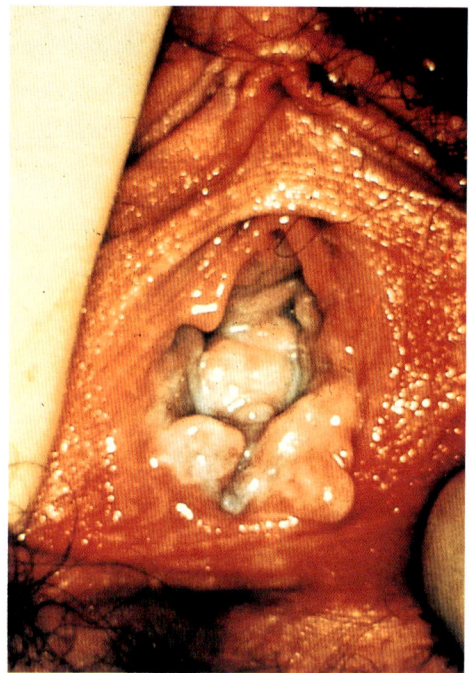

Figure 90.
Bacterial vaginosis with typical homogeneous vaginal discharge with foul odor.

Selected References

1. Heineman HS, Braude AI. Anaerobic infection of the brain. Observations on eighteen consecutive cases of brain abscess. *Am. J. Med.* 35:682-697, 1963.

2. Yoshikawa TT, Chow AW, Guze LB. Role of anaerobic bacteria in subdural empyema—report of four cases and review of 327 cases from the English literature. *Am. J. Med.* 58:99-104, 1975.

3. Chow AW, Roser SM, Brady FA. Orofacial odontogenic infections. *Ann. Intern. Med.* 88:392-402, 1978.

4. Mulligan ME. Ear, nose, throat, and head and neck infections. In Finegold SM, George WL (eds.), *Anaerobic Infections in Humans,* Academic Press, San Diego, 1989.

5. Felner JM, Dowell VR Jr. *"Bacteroides"* bacteremia. *Am. J. Med.* 50:787-796, 1971.

6. Felner JM, Dowell VR Jr. Anaerobic bacterial endocarditis. *N. Engl. J. Med.* 283:1188-1192, 1970.

7. Bartlett JG, Finegold SM, Anaerobic infections of the lung and pleural space. *Am. Rev. Respir. Dis.* 110:56-77, 1974.

8. Sabbaj J, Sutter VL, Finegold SM. Anaerobic pyogenic liver abscess. *Ann. Intern. Med.* 77:629-638, 1972.

9. Dunn DL, Simmons RL. The role of anaerobic bacteria in intraabdominal infection. *Rev. Infect. Dis.* 6:S139-S147, 1984.

10. Bennion RS, Baron EJ, Thompson Jr, JE, Downes J, Summanen P, Finegold SM. The bacteriology of gangrenous and perforated appendicitis—revisited. *Ann. Surg.* 211:165-171, 1990.

11. Chow AW, Marshall JR, Guze LB. Anaerobic infections of the female genital tract. Prospects and perspectives. *Obstet. Gynecol.* 30:477-494, 1975.

12. Goldstein EJC, Citron DM, Finegold SM, Role of anaerobic bacteria in bite wound infections. *Rev. Infect. Dis.* 6:S177-S183, 1984.

13. George WL, Other infections of skin, soft tissue, and muscle. In Finegold SM, George WL (eds.), *Anaerobic Infections in Humans.* Academic Press, San Diego, 1989.

14. Lewis RP, Sutter VL, Finegold SM. Bone infections involving anaerobic bacteria. *Medicine* (Balt.) 57:279-305, 1978.

15. Rolfe RD, Finegold SM (eds.), *Clostridium difficile: Its Role in Intestinal Disease.* Academic Press, Orlando, FL, 1988.

SECTION 3

SPECIMEN COLLECTION AND TRANSPORT

SECTION 3

SPECIMEN COLLECTION AND TRANSPORT

Introduction

The special atmospheric conditions, media, time, and cost required for inoculation and interpretation of anaerobic cultures demands that only specimens likely to yield clinically relevant results be processed by clinical laboratories.

The single most important factor in obtaining clinically relevant microbiology culture results is the quality of the specimen. Each specimen must be collected to avoid contamination by air and by normal mucosal flora (predominantly anaerobic bacteria) and the specimen must reach the laboratory in approximately the same condition.

Specimens obtained from a closed infectious process by aspiration through non-contaminated tissue are the most desirable and most likely to yield reliable culture results. Those obtained from infected fluids and tissues during surgical procedures, if placed under anaerobic conditions immediately, are also excellent.

Specimens obtained from sites with any indigenous flora, such as skin surface wounds, respiratory specimens obtained via the mouth (sputum, bronchial washings, tracheal suction, etc.), the vaginal vault, the gastrointestinal tract, and others are unacceptable for anaerobic culture under any circumstances.

The following pages illustrate the collection methods (pages 96-100) and transport methods (pages 101-105) that we have found work best for maximum anaerobic recovery. The Table (pages 106-110) lists the infectious processes introduced earlier and describes the appropriate specimens and handling procedures required to obtain meaningful microbiological information.

Specimen Collection

Aspiration of draining sinus tract or deep wound

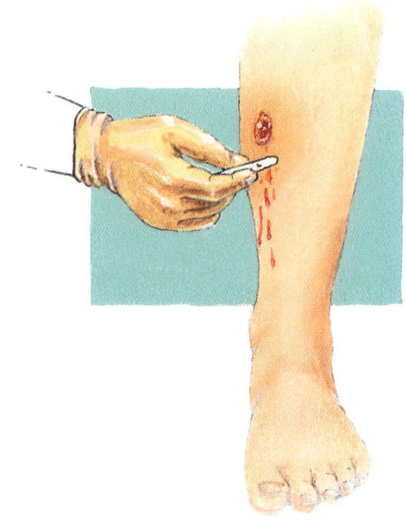

STEP 1: Thoroughly cleanse skin surface surrounding wound with povidone-iodine. Allow to remain wet for 1 min. Wash with ethanol to remove iodine. Allow ethanol to dry.

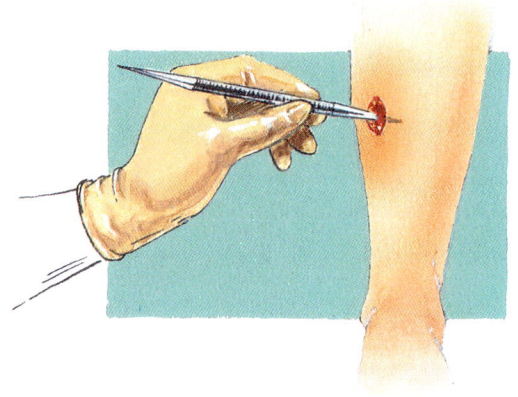

STEP 2: Obtain curetting from deep interior of wound or sinus tract.

NOTE: Swabs are not recommended because they are unlikely to yield clinically significant anaerobes.

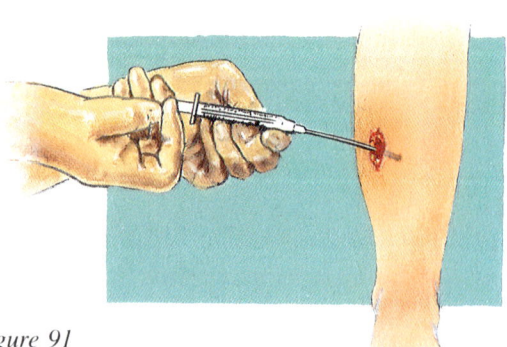

ALTERNATIVE METHOD: Obtain exudate from deep interior of wound using a plastic catheter and syringe.

Figure 91

Aspiration of subcutaneous abscess

STEP 1: Cleanse adjacent uninvolved skin with povidone-iodine. Allow to remain wet for 1 min. Wash with ethanol to remove iodine. Allow ethanol to dry.

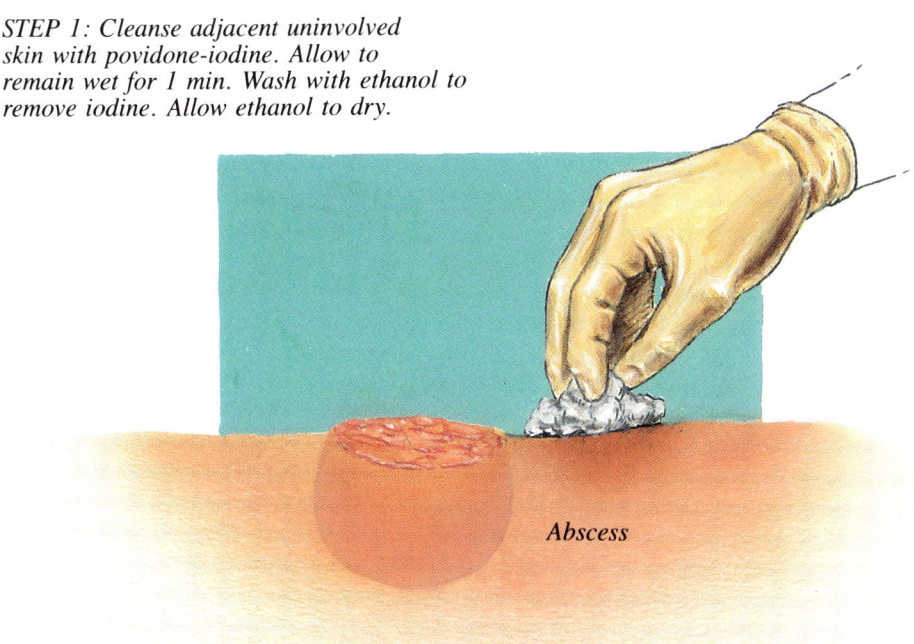

Abscess

STEP 2: Aspirate material through disinfected skin.

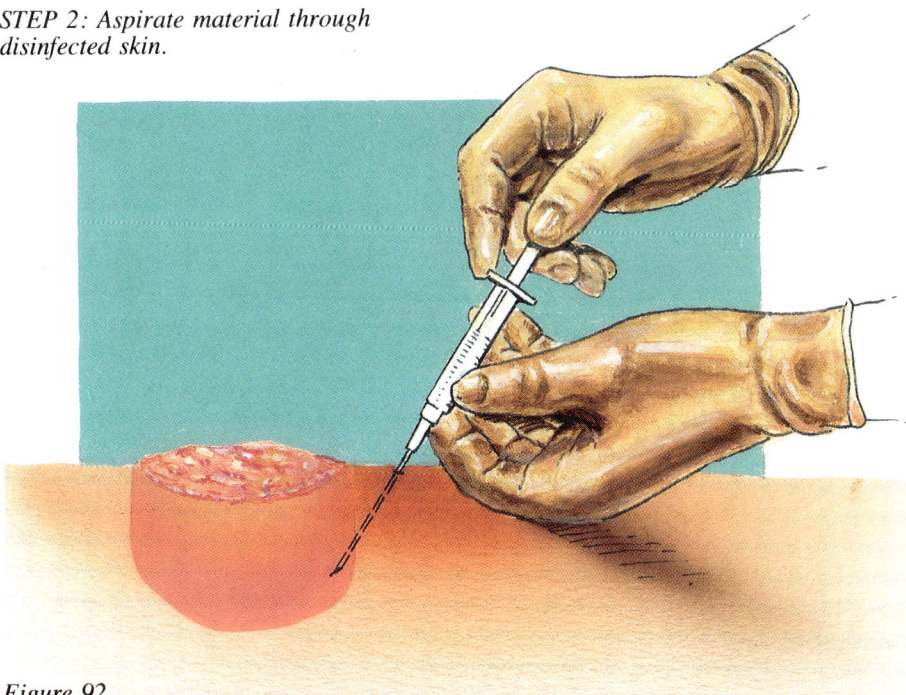

Figure 92

Obtaining specimen from contaminated surface wound or ulcer when aspiration is not possible.

STEP 1: Prepare wound by cleansing with wet-to-dry dressings over several days.

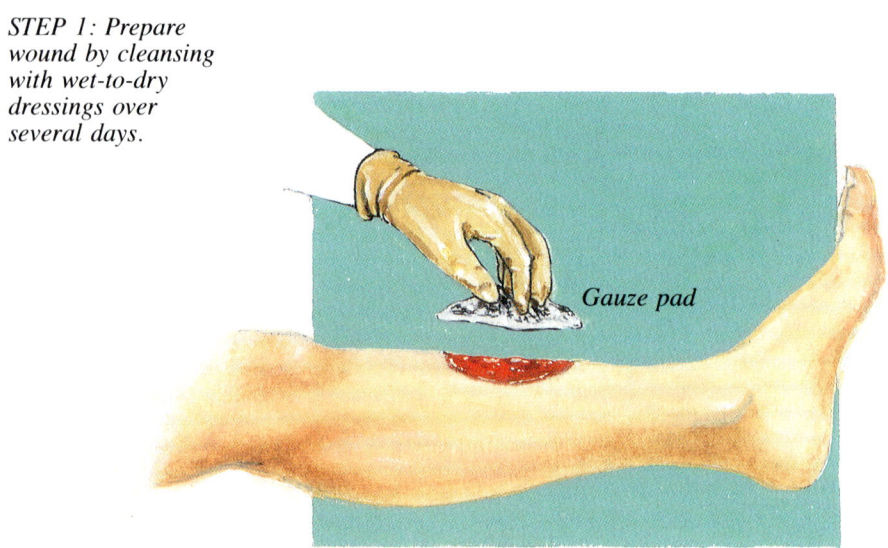

Gauze pad

STEP 2: Flush wound with 5 liters of povidone-iodine/sterile saline (50:50 solution) using gravity feed or pulsed jet.

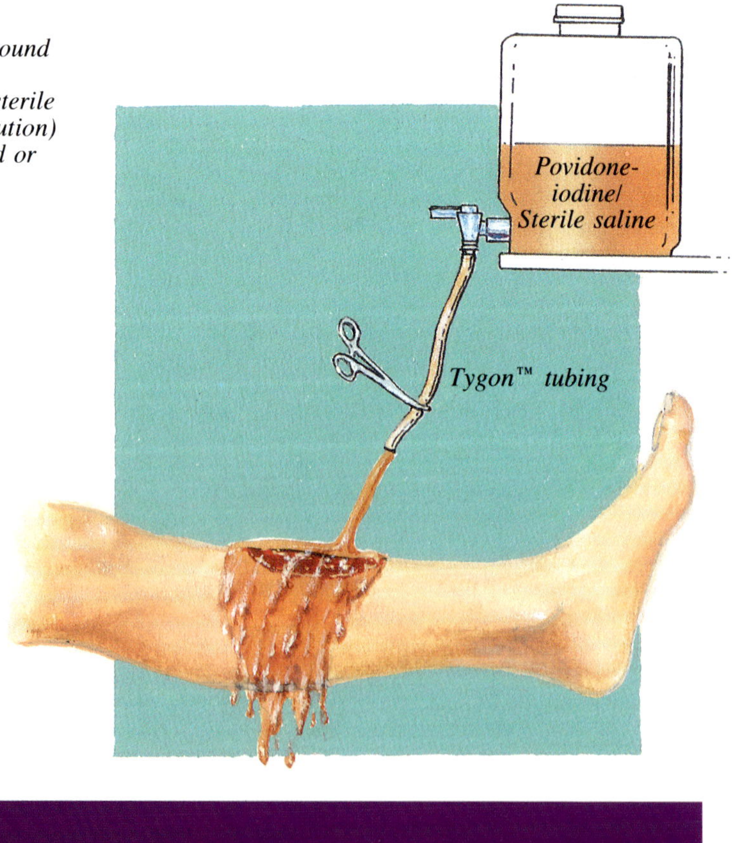

Povidone-iodine/ Sterile saline

Tygon™ tubing

Figure 93

STEP 3: Flush iodine from the wound with 5 liters of sterile saline using gravity feed or pulsed jet.

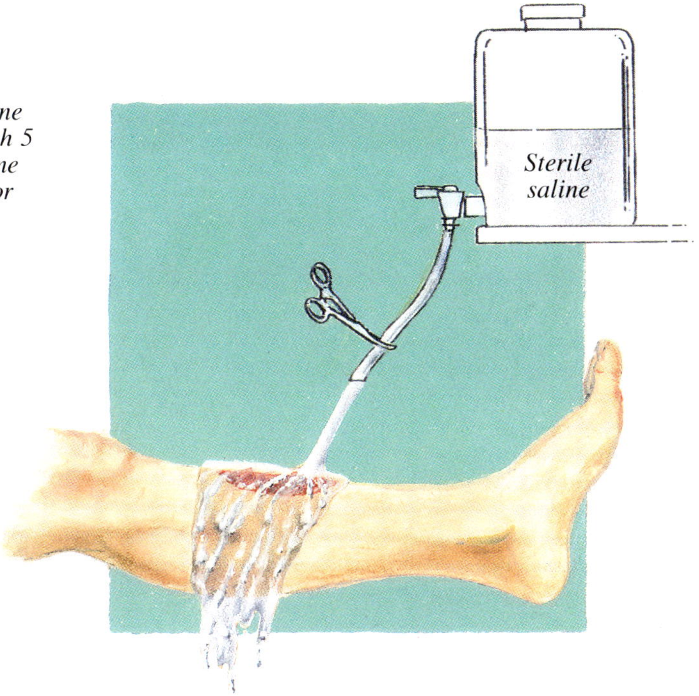

Sterile saline

STEP 4:
Use curette to obtain tissue from base of decontaminated ulcer.

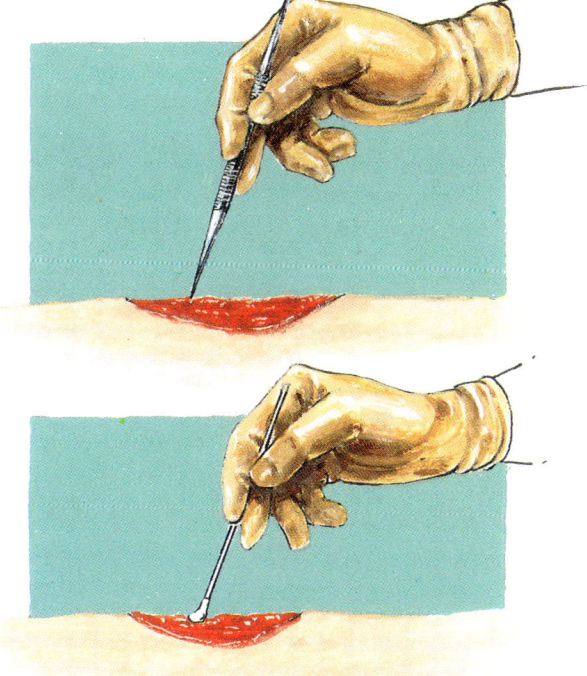

LESS RECOMMENDED ALTERNATIVE: Use swab to obtain specimen from base of decontaminated ulcer, avoiding contact with skin.

Figure 93

Transtracheal aspiration

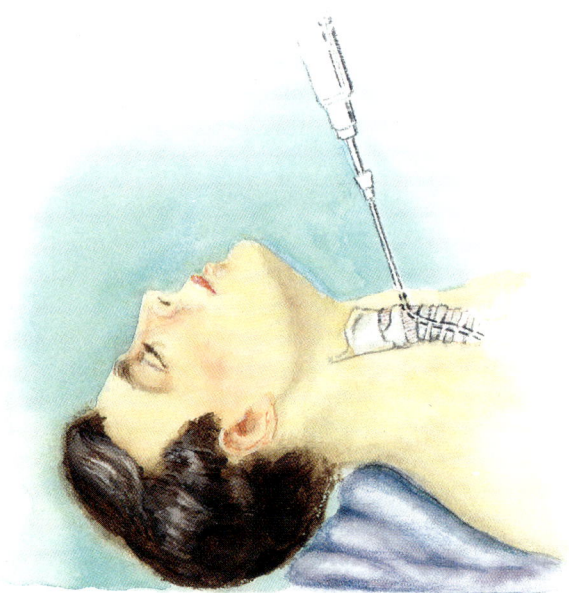

Figure 94.
Transtracheal aspiration procedure.

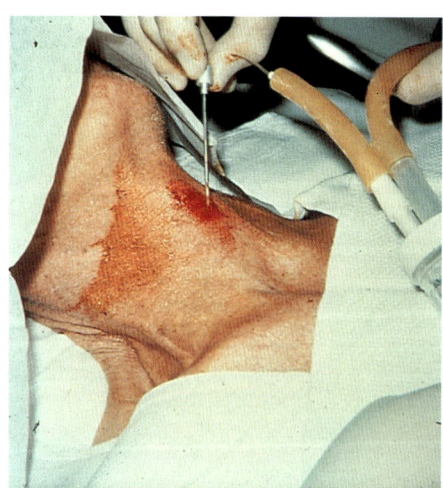

Figure 95.
Transtracheal aspiration (TTA).
Needle has been withdrawn, leaving polyethylene catheter in place; other end of
catheter is linked with Lukens trap, in turn linked with suction machine (not shown).

Specimen Transport

Anaerobic transport vial

The anaerobic transport vial is an ideal system for transporting aspirated specimens.

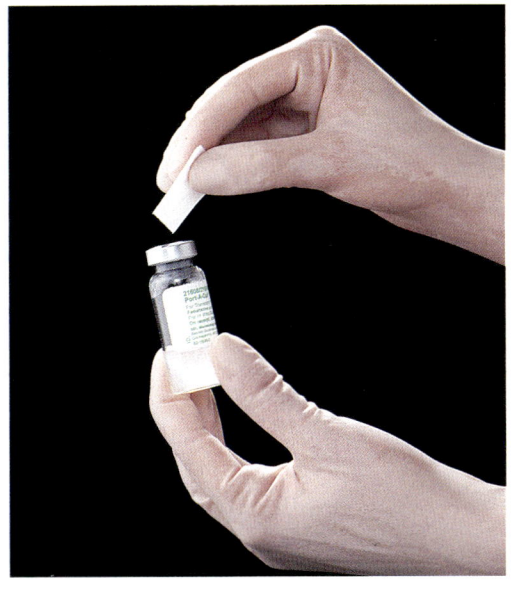

Figure 96.
STEP 1. Wipe septum with alcohol.

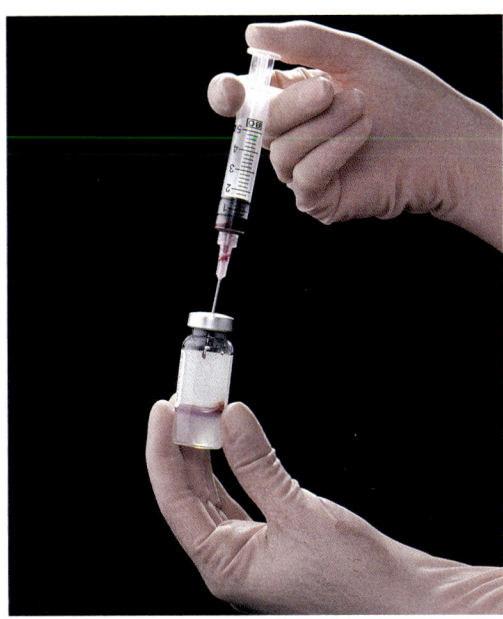

Figure 97.
STEP 2. Inject specimen without
introducing air into the
vial.
[Caution: inject slowly to
ensure that specimen
remains on top of agar]

Anaerobic pouch

The anaerobic pouch* is an ideal system for transporting tissue specimens.

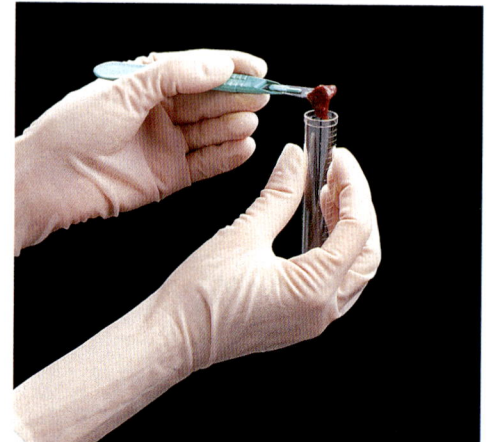

Figure 98.
STEP 1. *Place tissue aseptically into wide-mouth sterile plastic snap-cap tube.*

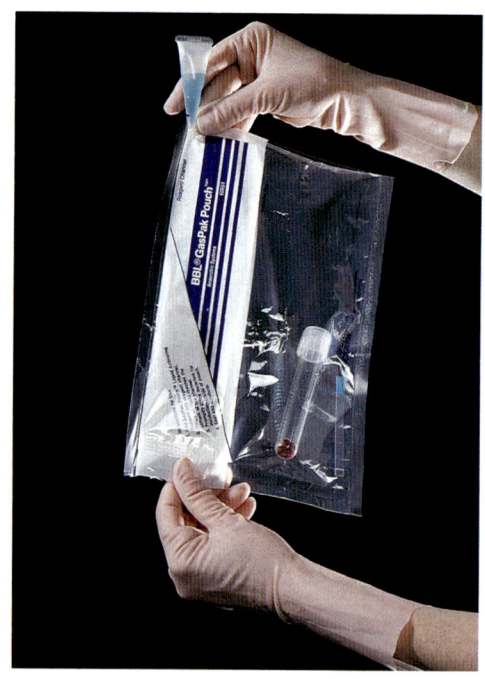

Figure 99.
STEP 2. *Place tube into pouch*
STEP 3. *Squeeze liquid from vial into channel in pouch and work liquid down onto catalyst.*

*BBL GasPak Pouch (Becton-Dickinson Microbiology Systems, Inc.)

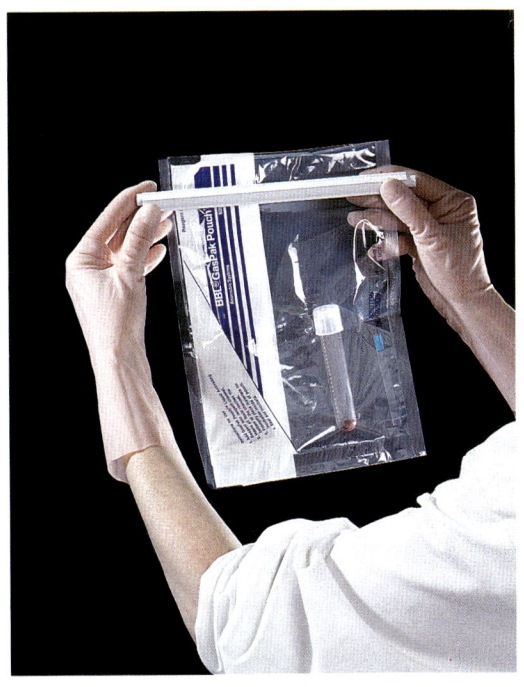

Figure 100.
STEP 4. Seal pouch with plastic rods.

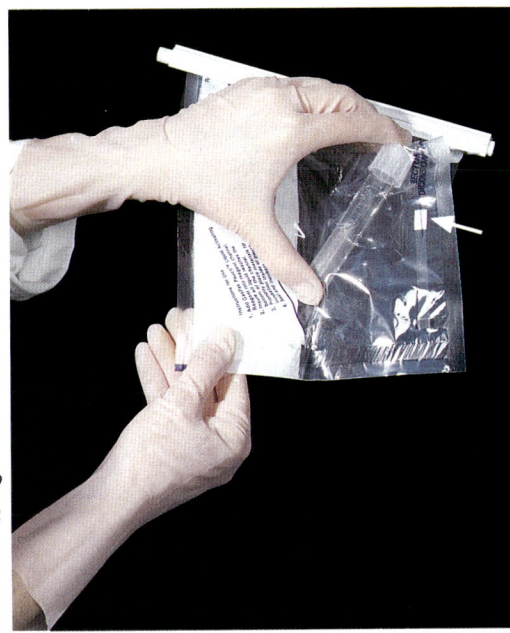

Figure 101.
STEP 5. When indicator strip has turned white, snap cap of tube tight to prevent desiccation. Indicator strip (arrow) color change from blue to white indicates anaerobic conditions within pouch.

Anaerobic transport tube

The anaerobic transport tube is a system for transporting swab and small tissue specimens.

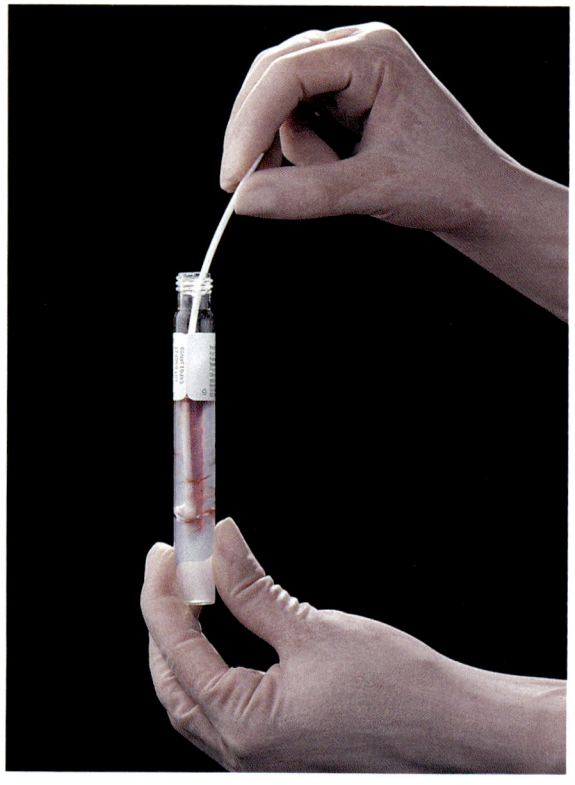

Figure 102.
STEP 1. Insert swab deep into agar (contains reducing substances to remove residual oxygen).

STEP 2. Break off contaminated end of swab shaft.

STEP 3. Immediately replace cap.

Alternative transport products

Figure 103.
Note: All use Hungate-type stoppers.
A. Tube on left is gassed-out tube containing sterile swab.
B. Center tube is a transport tube with screw cap incorporating a rubber septum. Specimens may be injected through the septum or the cap may be removed for insertion of swab or tissue.*
C. Tube on right is a small conical vial containing glass beads; appropriate for small-volume aspirates.

A   B   C

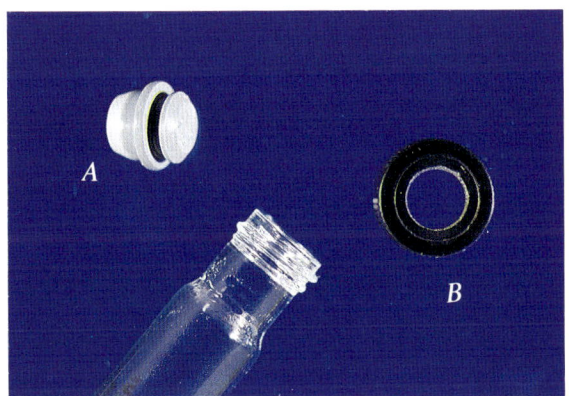

Figure 104.
Hungate-type stopper. Butyl rubber seal (A) prevents oxygen diffusion. Black plastic cap (B) secures stopper.

*Anaerobe Systems, San Jose, CA

Table 7. SUMMARY OF SPECIMEN COLLECTION AND TRANSPORT FOR ANAEROBIC CULTURE

Infectious Process	Specimen	Collection Method	Handling
Abscess Anorectal Buttock Breast Endodontal (See Endodontal abscess) Hidradenitis suppurativa Liver Lung Oral; orofacial Pilonidal	Abscess contents	Needle and syringe aspirate through skin decontaminated with povidone-iodine.	Anaerobic Transport Vial (ATV)
Brain abscess	Abscess fluid	Needle and syringe if sufficient material; Swab (least desirable)	ATV Anaerobic Transport Tube (ATT)
Abscess (viscera) or peritonitis	Abscess contents or fluid obtained during surgery	Use needle and syringe to aspirate fluid from loculated abscess areas. Sterile syringe alone without needle or with cannula can be used to collect fluid in body cavities. Avoid contact with mucosal or oro-gastrointestinal tract lumens. If insufficient fluid to aspirate, use swab.	Change to sterile needle before injecting material into ATV. ATT with agar butt
Abortion, septic	Blood (30 ml minimum) Uterine contents	Needle and syringe or blood collection tubing unit Sleeved endometrial culture curette, endometrial suction curette, or modified triple-lumen catheter. Tissue and fluid from high in uterus most important for relevant culture results.[1] (Also see Endometritis, below.)	Anaerobic blood culture bottle ATT
Actinomycosis (See Wound, deep)			
Appendicitis (CULTURE FOR RESEARCH PURPOSES ONLY)	Peritoneal fluid (purulent)	Syringe	ATV
	Appendix tissue obtained at surgery	Obtain tangential slice of appendix to avoid lumen	Petri dish or tube in anaerobic pouch

Infectious Process	Specimen	Collection Method	Handling
Arthritis, purulent (See Subcutaneous or Synovial fluid)			
Aspiration pneumonia	Transtracheal aspirate or protected bronchial brush catheter specimen	Catheter and syringe or bronchial brush.[2]	ATV
Bacteremia	Blood (30 ml minimum)	Needle and syringe or blood collection tubing unit	Sterile tube with SPS or anaerobic blood culture bottle
Bacterial vaginosis (CULTURE FOR RESEARCH PURPOSES ONLY)	Vaginal discharge	Collect swab of material in posterior fornix of vagina	Gram stain; examine for decreased lactobacilli and numerous gram-variable coccobacilli and curved rods. Examine fluid for pH > 4.5, amine odor with KOH, clue cells on wet preparation, and homogeneous gray discharge.
Biliary tract infection (See Abscess, visceral)			
Bite wound infection (See Wound, deep)			
Botulism (Note: Notify Public Health officials and hold all materials.) Use care; These materials are dangerous!	Serum	Vacutainer or needle and syringe	Red-top tube. Remove serum from clot and refrigerate.
	Feces (> 50 g), vomitus or gastric aspirate		Refrigerate in sterile urine cup.
	Implicated food		Refrigerate in original container.
Botulism, wound	Exudate	Swab	ATT
	Tissue biopsy or curettings		Sterile petri dish or tube in anaerobic pouch
Bowel bacterial overgrowth syndrome (RESEARCH PURPOSES ONLY)	Small bowel lumen contents and blind loop pouch contents	Collect material through special tube inserted via the nose.	ATV or place material into sterile tube in anaerobic pouch.
Cellulitis (Note: Productive in rare cases)	Aspirate from leading edge of lesion	Needle and syringe; obtain through disinfected skin. May require injection of nonbacteriostatic saline.	ATV
Chorioamnionitis	Exposed surface of extraplacental membrane	Separate chorion and amnion of placenta. Obtain swabs from extraplacental membrane surface[3]	ATT

Infectious Process	Specimen	Collection Method	Handling
Diverticulitis (RESEARCH PURPOSES ONLY)	Abscess contents	Drain material under ultrasound, CT or at surgery	ATV
Drainage, wound (See Wound, deep)			
Empyema (subdural, thoracic, gallbladder)	Fluid or aspirated pus	Needle and syringe or catheter and syringe, or at surgery	Change to sterile needle before injecting material into ATV.
Endocarditis	Blood (> 30 ml)	Needle and syringe or collection system	As for Bacteremia
	Heart valve tissue	Sterile container	As for other tissue
Endodontal abscess (RESEARCH PURPOSES ONLY)	Aspirate from base of tooth	Needle and syringe; isolate tooth with rubber dam; decontaminate with povidone-iodine, dry, drill into root canal, aspirate pus.	ATV
Endometritis	Intrauterine tissue and fluid	Obtain intrauterine contents with sampling curette protected by plastic sleeve (Pipelle® endometrial suction curette, etc.)[1]	ATV
Enteritis necroticans (CULTURE FOR RESEARCH PURPOSES OR BY PUBLIC HEALTH LABORATORY)	Feces		Sterile container. Examine Gram stain for predominance of characteristic clostridial morphotypes.
Gangrene (See Tissue infection)			
Myonecrosis (See Tissue infection)			
Necrotizing fasciitis (See Tissue infection)			
Oral infection Abscess (See Abscess) Endodontal abscess (See Endodontal abscess)			

Infectious Process	Specimen	Collection Method	Handling
Osteomyelitis	Bone biopsy	Obtain surgically, avoid contamination with infected surface tissue.	Petri dish or plastic tube in anaerobic pouch
	If unable to biopsy bone, obtain curetting or aspirate from deep site.	Decontaminate surface with povidone-iodine, debride necrotic tissue, obtain material from base of wound.	
Otitis media	Aspirate from middle ear (by physician)	Tympanocentesis through eardrum decontaminated with povidone-iodine	ATV
		If insufficient fluid to aspirate, obtain swab after decontaminating site with iodine lavage followed by sterile saline lavage.	ATT
Periodontal infection	Sterile paper points or wire broach.	Isolate site with rubber dams, decontaminate surface with povidone-iodine, insert point or broach into pocket[4]	ATV
Peritonitis (See Abscess, viscera)			
Salpingitis (See Endometritis or Abscess, viscera)			
Sinus tract (See Wound, deep) Actinomycosis Other draining wounds			
Sinusitis, chronic	Aspirated pus or washing	Disinfect nasal cavity below anterior part of inferior turbinate with 2% tetracaine and wait 20 min. Insert sterile 2 mm puncture needle into maxillary antrum and aspirate material into 20 ml syringe. If no material obtained, inject 1-2 ml sterile nonbacteriostatic saline and aspirate washings.[5]	ATV
Subcutaneous or synovial infection	Fluid aspirate	Needle and syringe; aspirate through disinfected skin.	ATV
Tetanus (NOTE: NOTIFY PUBLIC HEALTH AUTHORITIES AND HOLD ALL MATERIALS).	Serum	Red-top tube	Pour serum off clot and refrigerate.
	Aspirated pus	Needle and syringe, catheter and syringe, or swab.	ATV or ATT

Infectious Process	Specimen	Collection Method	Handling
Ulcer (decubitus, diabetic, Meleney's)	Tissue from base of ulcer. (SURFACE SWAB NOT ACCEPTABLE FOR CULTURE)	Clean and debride surface with wet-to-dry dressings over several days prior to obtaining specimen. Flush wound surface with 5 liters of 50% povidone-iodine in sterile saline followed by 5 liters of sterile saline, applied by gravity-feed or pulsed-jet spray. Obtain curetings from base of ulcer.	Petri dish or tube in anaerobic pouch.
Urinary tract, bacteriuria	Bladder urine (RARELY INDICATED)	Obtain by suprapubic bladder aspiration with needle and syringe through disinfected skin.	ATV
	Blood (30 ml minimum)	Needle and syringe or blood collection tubing unit	Anaerobic blood culture bottle
Vincent's angina (CULTURE FOR RESEARCH PURPOSES ONLY)	Material from lesion	Swab tonsillar surface or ulcer.	Gram stain lesion material and examine for characteristic fusiforms.
Wound, deep	Tissue from base of wound	Decontaminate skin surface with povidone-iodine, insert curette deep into base of wound and obtain tissue.	Petri dish or tube in anaerobic pouch
	Aspirated pus	Thread plastic catheter deep into wound or sinus tract and aspirate into syringe.	ATV
		As last choice, insert swab deep into wound after thorough skin decontamination.	ATT

Selected References

1. Pipelle® endometrial suction curette, Unimar Incorporated, Wilton, CT (described in Martens MG, Faro S, Hammill HA, Riddle GD, Smith D. Transcervical uterine cultures with a new endometrial suction curette: a comparison of three sampling methods in postpartum endometritis. *Obstet. Gynecol.* 74:273-276, 1989; or Endometrial culture curette, Rocket Corp. Contact Doris C. Brooker, M.D., Univ. Minnesota (612) 626-5001. Also see Eschenbach DA, Rosene K, Tompkins LS, Watkins H, Gravett MG. Endometrial cultures obtained by a triple-lumen method from afebrile and febrile postpartum women. *J. Infect. Dis.* 153:1038-1045, 1986.)

2. Broughton WA, Bass JB, Kirkpatrick MB. The technique of protected brush catheter bronchoscopy. *J. Crit. Illness* 2:63-70, 1987.

3. Hillier SL, Martius J, Krohn M, Kiviat N, Holmes KK, Eschenbach DA. A case-control study of chorioamnionic infection and histologic chorioamnionitis in prematurity. *N. Engl. J. Med.* 319:972-978, 1988.

4. Newman MG, Socransky SS. Predominant cultivable microbiota in periodontosis. *J. Periodont. Res.* 12:120-128, 1977.

5. Jousimies-Somer HR, Savolainen S, Ylinskoski JS. Bacteriological findings of acute maxillary sinusitis in young adults. *J. Clin. Microbiol.* 26:919-925, 1988.

SECTION 4

GRAM STAINS

SECTION 4

GRAM STAINS

Introduction

Gram Stains of Clinical Specimens

The direct Gram stain of clinical material often reveals clues to the anaerobic nature of an infectious process. Such Gram stains, performed immediately upon receipt of the specimen in the laboratory, provide rapid and relevant information to clinicians that can be used to direct initial therapy. Examples of Gram stains of material from some of the clinical entities discussed earlier are illustrated here. The descriptions are similar to those that would be reported by microbiology laboratory technologists.

All Gram stains were air-dried and fixed with 95% methanol for 1 minute. After the methanol had dried, the smears were stained. Methanol fixation results in less distortion of bacteria and better visualization of human cells than does heat fixation. Dilute carbol fuchsin was used in place of the standard safranin counter-stain to enhance visibility of gram-negative structures.

Gram Stains of Pure Cultures

Selected species of anaerobic organisms were grown in pure culture to illustrate key morphologic characteristics.

All organisms were grown for 48 hours in an anaerobic atmosphere on blood agar plates (supplemented with vitamin K). Material from one colony was emulsified in saline, air-dried, and fixed with methanol before staining (as above).

Clinical Specimens

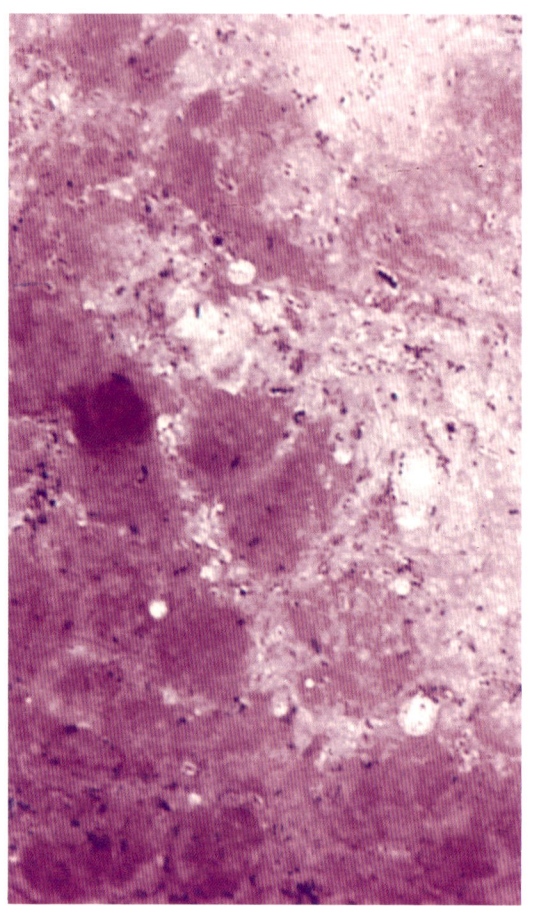

Figure 105.
Brain abscess. *Few PMNs and necrotic tissue debris are surrounded by numerous gram-positive cocci (singly and in short chains), gram-positive bacilli, and gram-negative pleomorphic bacilli and coccobacilli.*

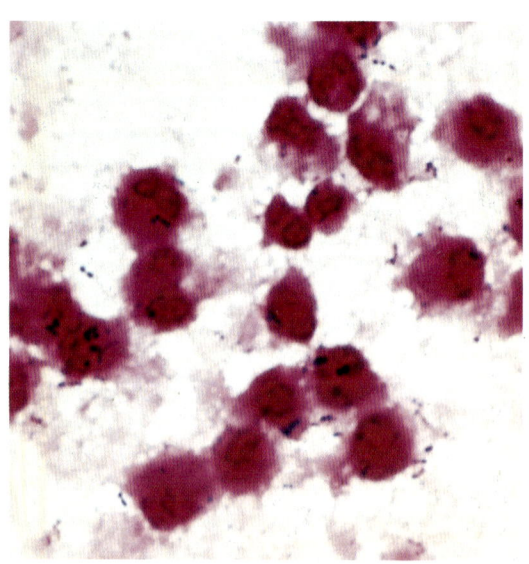

Figure 106.
Pleural effusion. *Numerous PMNs are surrounded by gram-positive cocci in chains and some small, pale gram-negative bacilli.*

Figure 107.
Empyema fluid *containing very few PMNs. Confluent bacterial morphotypes include gram-positive cocci in chains and pairs, and pale, pleomorphic gram-negative bacilli and coccobacilli.*

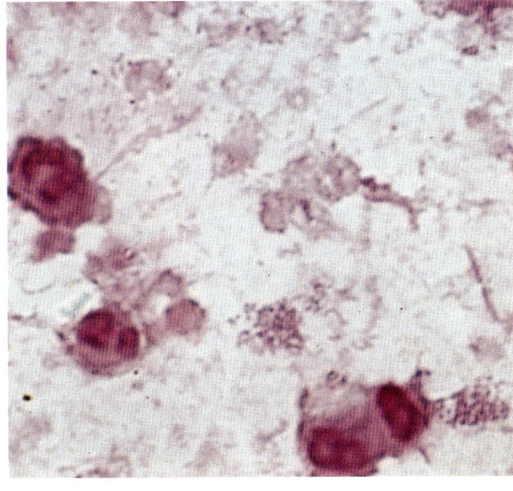

Figure 108.
Transtracheal aspirate. *Moderate PMNs and numerous small, gram-negative coccobacilli and a few pleomorphic gram-negative bacilli are seen.*

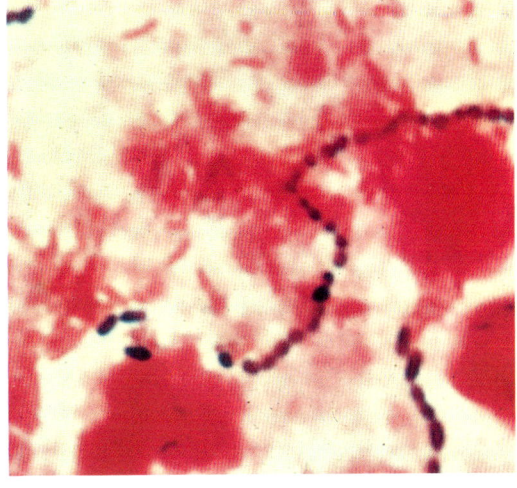

Figure 109.
Transtracheal aspirate.
Numerous PMNs, numerous gram-positive cocci in chains, and numerous regular-shaped gram-negative bacilli.

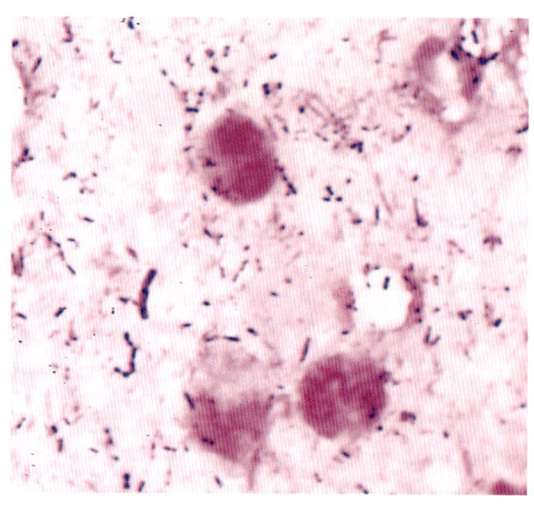

Figure 110.
Peritoneal fluid. *Numerous PMNs, numerous gram-negative bacilli and coccobacilli, numerous gram-positive cocci, and numerous gram-positive bacilli (ranging from pleomorphic to those resembling clostridia). This specimen is indicative of mixed fecal flora; cultures yield an average of 12 species per specimen.*

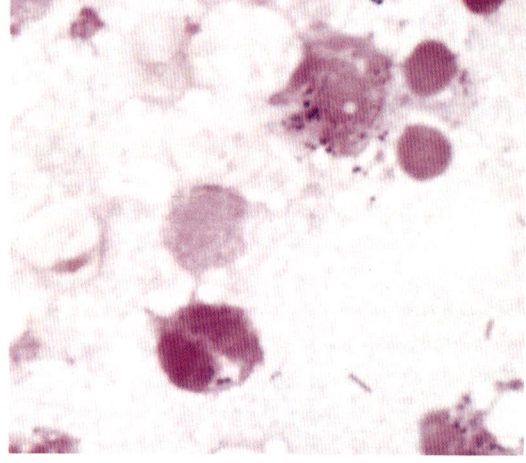

Figure 111.
Peritoneal fluid *showing numerous PMNs and numerous RBC ghosts. Bacteria include small gram-positive cocci in pairs and clusters (mostly intracellular), pleomorphic gram-positive bacilli, and small gram-negative bacilli and coccobacilli.*

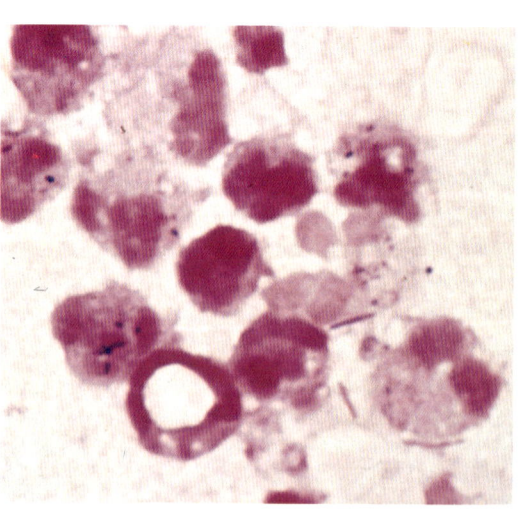

Figure 112.
Peritoneal fluid *showing numerous PMNs and few RBC ghosts. Bacteria include small gram-positive cocci in pairs and clusters (mostly intracellular), pleomorphic gram-positive bacilli, a few large, pleomorphic gram-negative bacilli, and small gram-negative coccobacilli.*

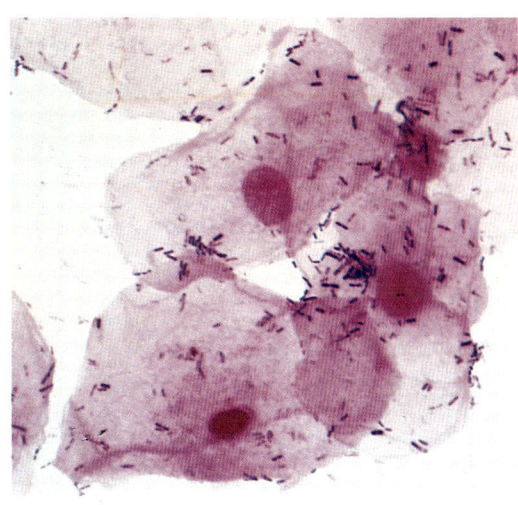

Figure 113.
Normal vaginal secretions.
Squamous epithelial cells with clearly defined borders and numerous large, gram-positive, parallel-sided bacilli (indicative of Lactobacillus *species) are prominent features.*

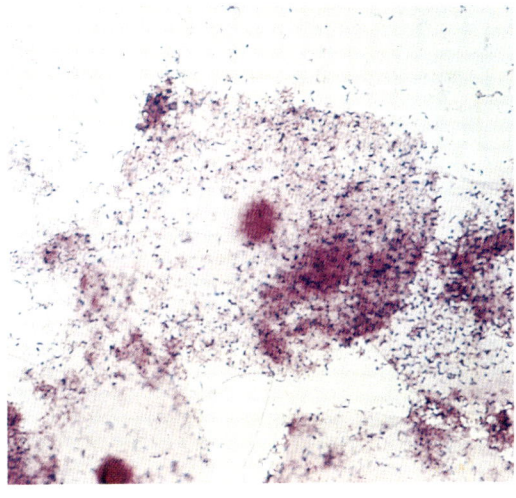

Figure 114.
Vaginal secretions from patient with bacterial vaginosis.
Epithelial cell borders are obscured by numerous small, coccobacillary gram-variable bacteria. Pleomorphic filamentous gram-negative bacilli are also present in this example. Of great importance is the lack of any organisms resembling lactobacilli.

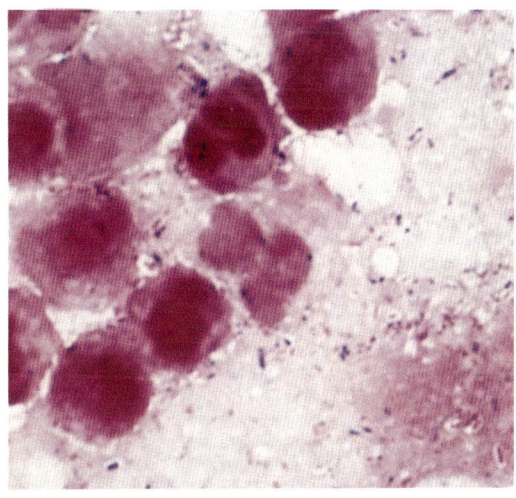

Figure 115.
Aspirate of intravenous drug abuser injection site abscess.
Smear shows numerous PMNs, numerous gram-positive cocci in pairs and chains, pleomorphic gram-positive bacilli, and gram-negative bacilli and coccobacilli (suggestive of oral flora-type organisms).

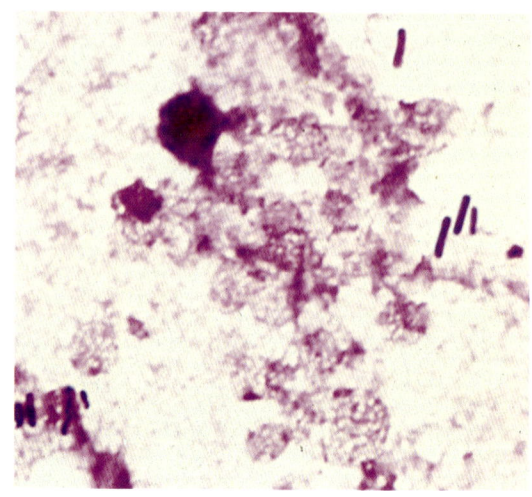

Figure 116.
Biopsy of gangrenous tissue from leg. *RBC ghosts, rare PMNs, and large, boxcar-shaped gram-positive and gram-variable rods without visible spores are prominent features of clostridial gangrene due to* Clostridium perfringens. *The organism elaborates a leukotoxin that degrades white cells, which accounts for the lack of PMNs in smears made from such specimens.*

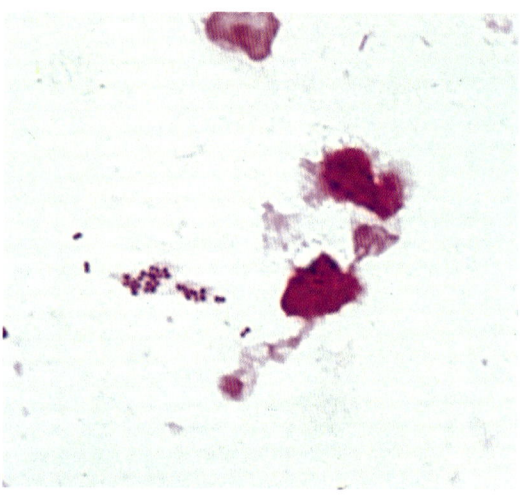

Figure 117.
Material from base of a diabetic foot ulcer. *Rare PMNs and epithelial cells are seen. Bacteria include gram-positive cocci in pairs and small clusters and pale, pleomorphic gram-negative bacilli.*

Figure 118.
Actinomycotic abscess, right cheek. *Smear shows numerous RBC ghosts, few PMNs, and large, pleomorphic, branching gram-positive bacilli (some intracellular) suggestive of* Actinomyces *species. The irregular staining seen here is not uncommon.*

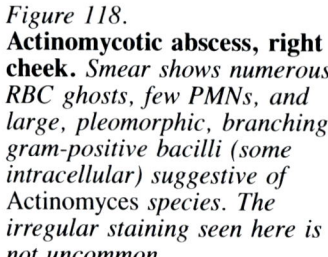

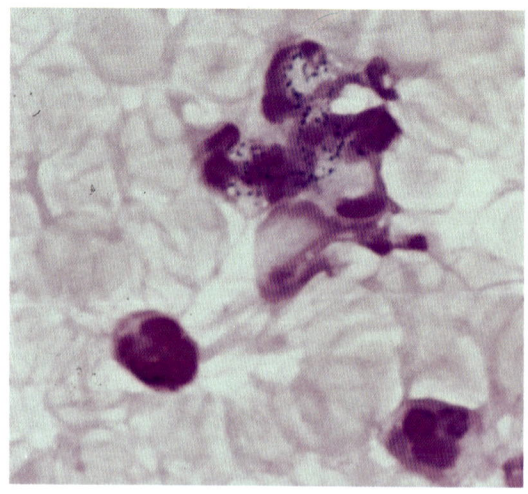

Pure Cultures

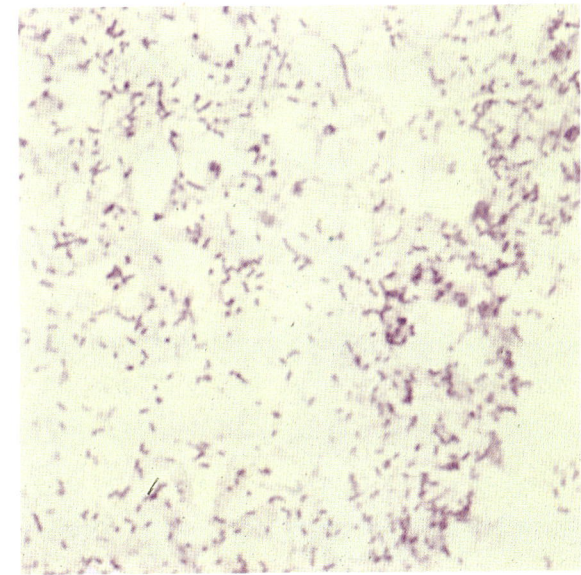

Figure 119.
Bacteroides buccalis.
Pleomorphic gram-negative bacilli and coccobacilli characterize this member of the normal oral flora.

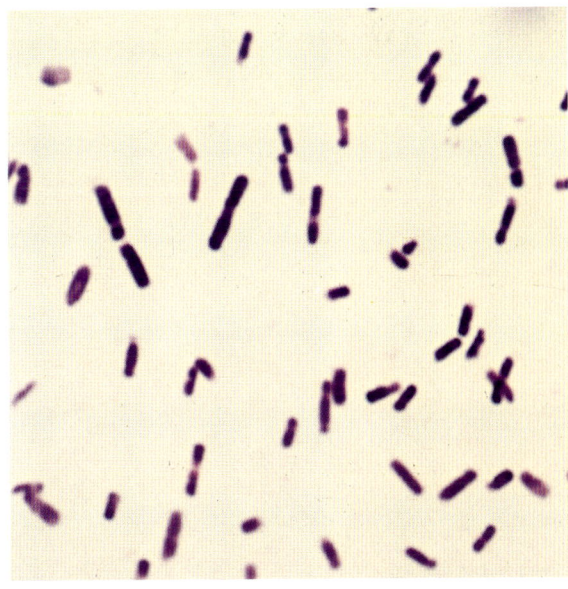

Figure 120.
Clostridium perfringens.
Large, gram-positive and easily decolorized (appearing gram-negative), square and boxcar-shaped rods without visible spores.

Figure 121.
Clostridium ramosum.
Thin, often curving, gram-positive and gram-variable rods with oval-to-round terminal spores.

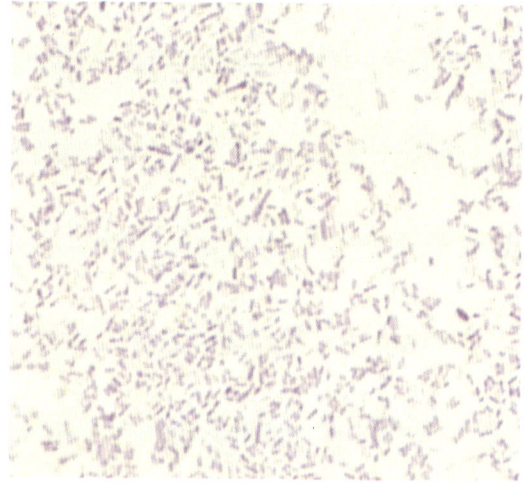

Figure 122.
Bacteroides melaninogenicus.
Small, pale, pleomorphic gram-negative bacilli and coccobacilli, sometimes displaying filamentous forms.

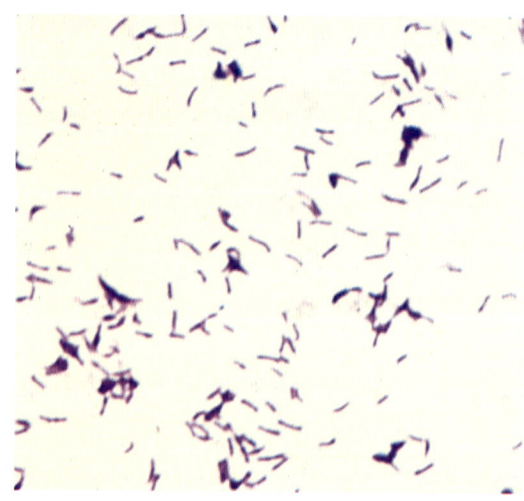

Figure 123.
Propionibacterium acnes.
Very pleomorphic gram-positive bacilli, showing coryneform morphology (clubbing, snapping forms, etc.). Called the "anaerobic diphtheroid."

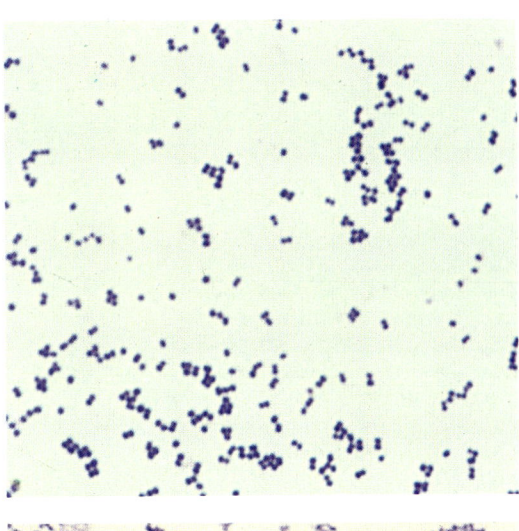

Figure 124.
Peptostreptococcus asaccharolyticus. *Medium-sized gram-positive cocci in pairs, short chains, and small clusters.*

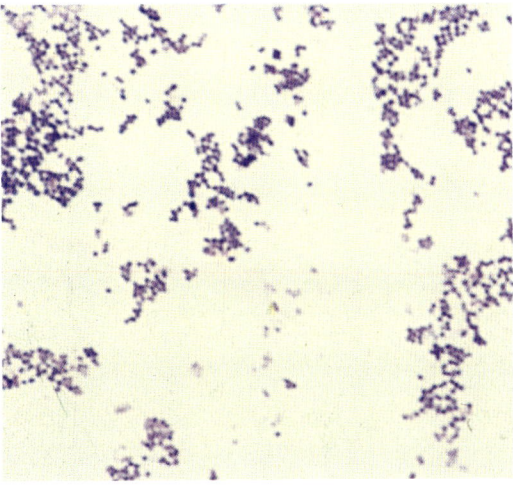

Figure 125.
Peptostreptococcus micros.
Tiny (< 0.6 μ diameter) gram-positive cocci in pairs and chains.

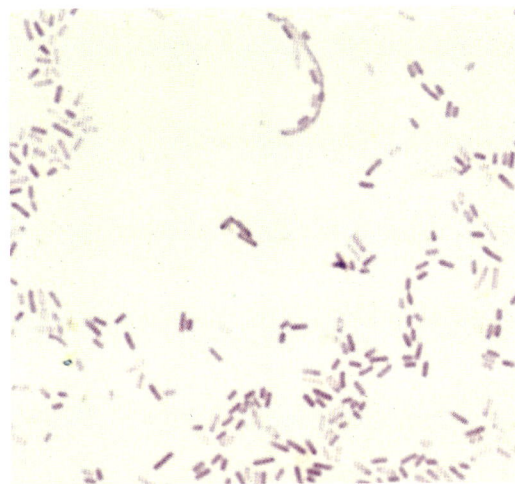

Figure 126.
Bacteroides fragilis.
Large, deeply staining gram-negative bacilli; often display filaments, other pleomorphisms, and vacuole-like paler-staining swellings within the cells. All members of the B. fragilis *group appear similar on Gram stain.*

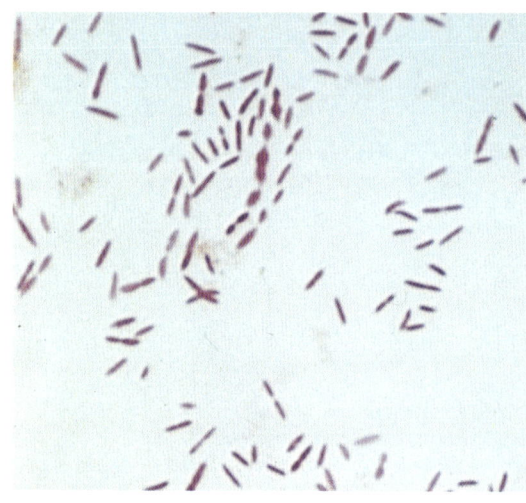

Figure 127.
Fusobacterium naviforme.
One example of the numerous morphologies exhibited by fusobacteria. This species is characterized by darkly staining, pleomorphic fusiform bacilli with pointed ends resembling small boats (from which the name is derived).

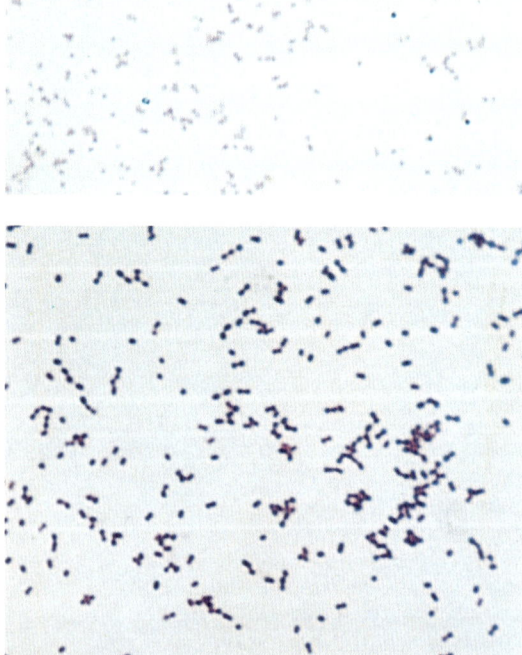

Figure 128.
Veillonella parvula.
Small, gram-negative diplococci whose morphology resembles that of the neisseriae. The pairs of cocci face each other parallel to their long axis, which distinguishes them from coccobacilli.

Figure 129.
Peptostreptococcus anaerobius.
Gram-positive cocci in pairs and chains; size similar to P. asaccharolyticus, *but* P. anaerobius *is more likely to be found in chains.*

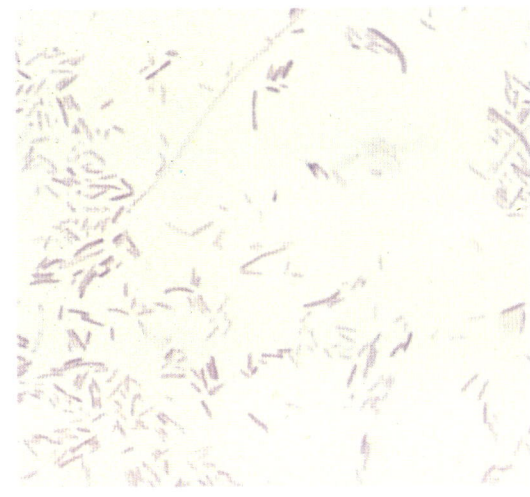

Figure 130.
Bilophila wadsworthia.
Pleomorphic, pale-staining gram-negative bacilli. Filamentous forms and B. fragilis-*like vacuoles are not uncommon. In the presence of the growth factor, bile, the organism exhibits a more regular morphology.*

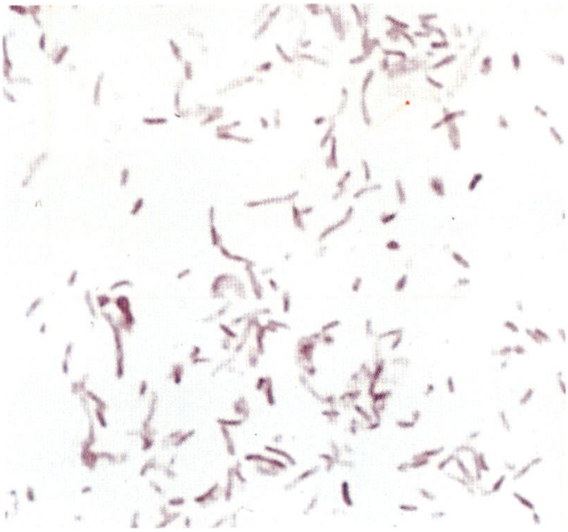

Figure 131.
Fusobacterium mortiferum.
Very irregular, pleomorphic gram-negative bacilli with prominent swellings and bulges.

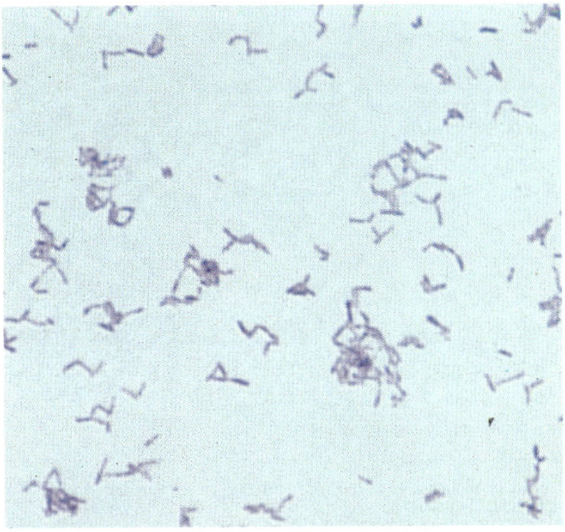

Figure 132.
Actinomyces israelii.
Gram-positive, branching and beaded pleomorphic bacilli. May exhibit short forms that resemble Propionibacterium acnes.

Figure 133.
Fusobacterium nucleatum.
Pale, fusiform, sharply pointed filamentous gram-negative bacilli.

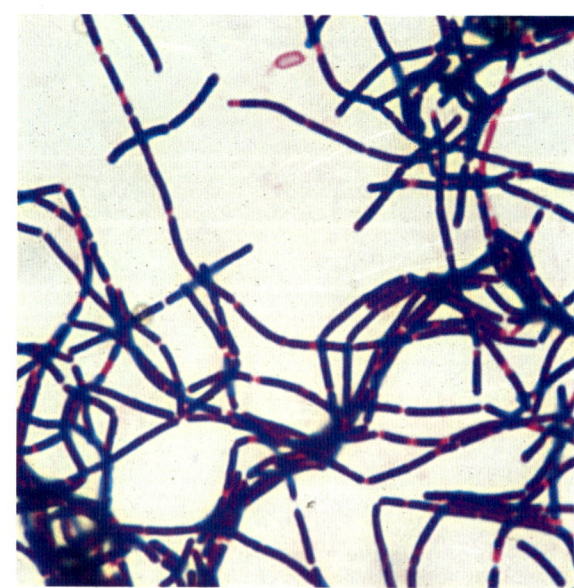

Figure 134.
Clostridium septicum.
Long, filamentous gram-positive bacilli with rare spores (seen as clear, oval-shaped objects) characterize this very motile anaerobe.

SECTION 5

SUSCEPTIBILITY OF ANAEROBES TO ANTIMICROBIAL AGENTS

SECTION 5

SUSCEPTIBILITY OF ANAEROBES TO ANTIMICROBIAL AGENTS

Susceptibility Testing Methods

A basic understanding of how antimicrobial susceptibility testing is carried out with anaerobes will assist the clinician in interpreting susceptibility data obtained from the laboratory. The advantages and disadvantages of the methods are listed, along with a step-by-step discussion of the testing method.

Susceptibility Testing is Recommended for:
Blood culture isolates
CNS and other serious infections
Anaerobes isolated in pure culture
Isolates from patients not responding to therapy

Commonly used Susceptibility Testing Methods for Anaerobes Include:
Agar dilution
Microbroth dilution
Broth-disk elution (This method has been commonly used and is easy to perform; however, serious questions have been raised about its reliability with many of the newer cephalosporins, with metronidazole, and with tetracycline. It is no longer approved by the National Committee for Clinical Laboratory Standards [NCCLS].)

The standardized Bauer-Kirby method used for aerobic susceptibility testing is not suitable for use in anaerobic testing.

Agar Dilution Technique

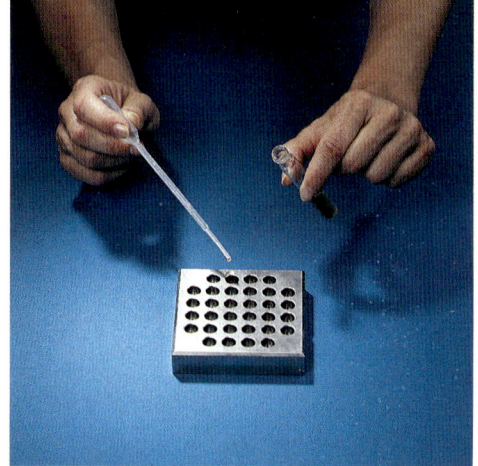

Figure 135.
Strains of various anaerobes are
added to the wells of an
inoculating device.

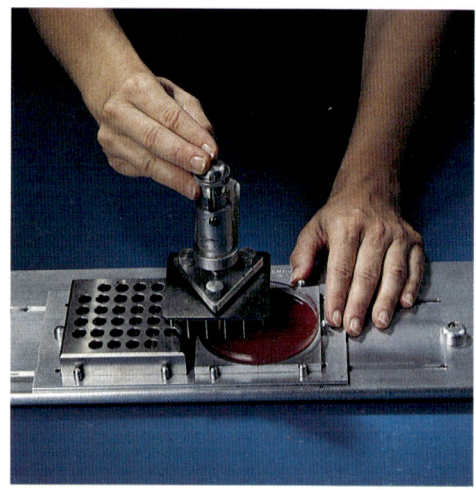

Figure 136.
Plates (containing various
concentrations of antibiotics) are
inoculated with these organisms.

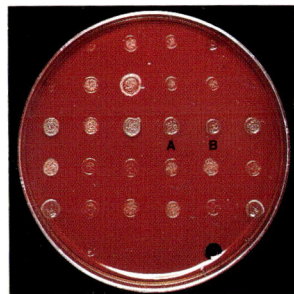

Figure 137.
  Plate 1. Growth
  Control.

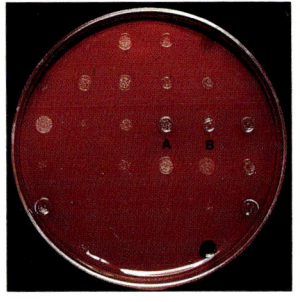

Figure 138.
  Plate 2. Antimicrobial
  Agent (16 μg/ml).

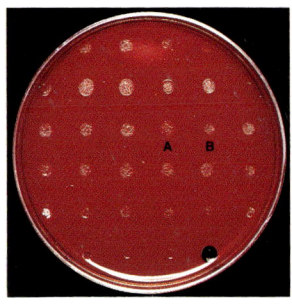

Figure 139.
  Plate 3. Antimicrobial
  Agent (32 μg/ml).

Each spot on the plate is a different strain. Minimum inhibitory concentration (MIC) is defined as the concentration of antibiotic at which there is no growth, one or two colonies, or a barely visible haze. In this example, strains A and B grow at 16 μg/ml (Plate 2) but not at 32 μg/ml (Plate 3); thus, the MIC for these strains is 32 μg/ml. (The residual spots seen at 32 μg/ml do not represent growth.) **Note that the nature of growth of different organisms may vary. MICs for organisms with weak or transparent growth on growth control, as well as MICs for organisms with "trailing endpoints," may be difficult to determine.**

Table 8. AGAR DILUTION

Advantages	Disadvantages
Many strains can be tested at one time	Cumbersome
	Labor-intensive
Supports the growth of even fastidious anaerobes	Not convenient for testing only a few strains
	Antibiotics must be decided upon in advance
	Minimum bactericidal concentrations (MBCs) cannot be determined

Microbroth Dilution Technique

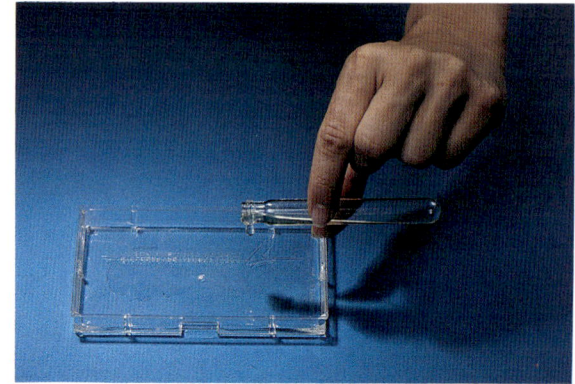

Figure 140.
The tube containing the strain to be tested is poured into a tray.

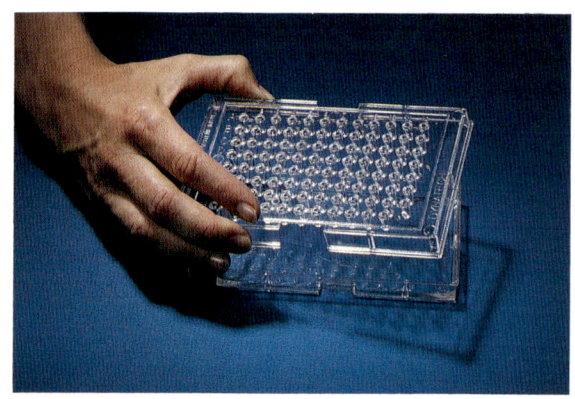

Figure 141.
A plastic multipoint inoculator is dipped into the tray containing the organism . . .

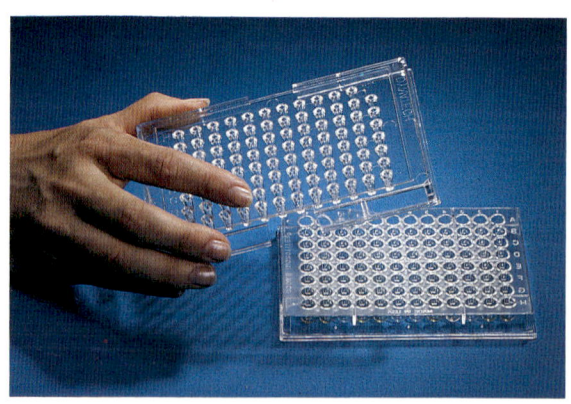

Figure 142.
And then into the microdilution tray, which contains serial dilutions of various antibiotics.

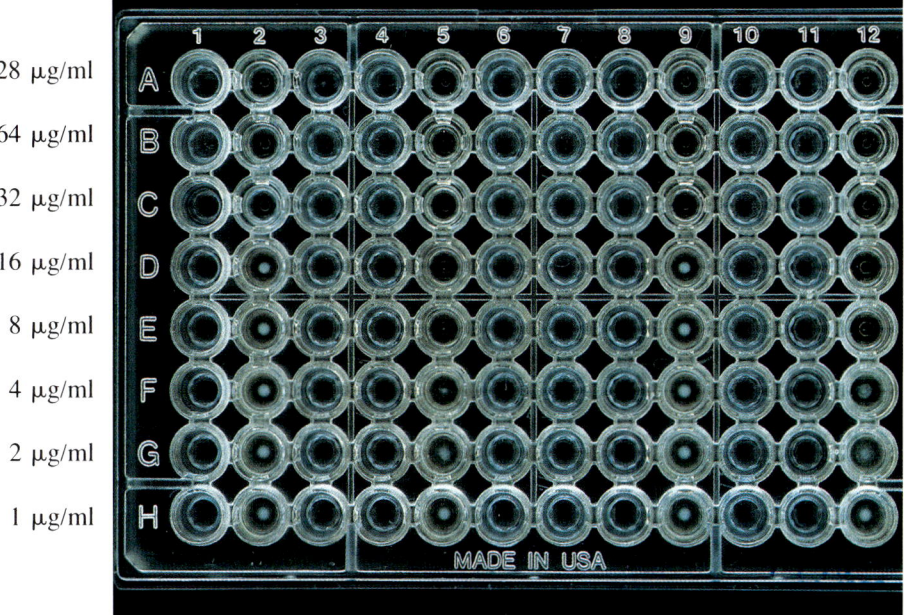

128 μg/ml

64 μg/ml

32 μg/ml

16 μg/ml

8 μg/ml

4 μg/ml

2 μg/ml

1 μg/ml

Figure 143.
The MIC is defined as the concentration in the first well showing no button of
growth. In this example, serial dilutions of antibiotics are placed in the columns
numbered 2, 5, 9, and 12. In column 2, growth is present at 16 μg/ml, but not at
32 μg/ml. The MIC for this organism is 32 μg/ml.

Table 9. MICROBROTH DILUTION

Advantages	Disadvantages
Plates can be prepared in advance and frozen or lyophilized, or purchased commercially	Does not sufficiently support the growth of many clinically relevant anaerobes (some *Fusobacterium*, *Bilophila*, *Peptostreptococcus*, pigmented *Bacteroides* and *Porphyromonas*)
Convenient for testing single isolates or many strains	
	"Trailing endpoints" may cause difficulty in MIC determination
	Not reliable for metronidazole

Broth Disk Elution Technique

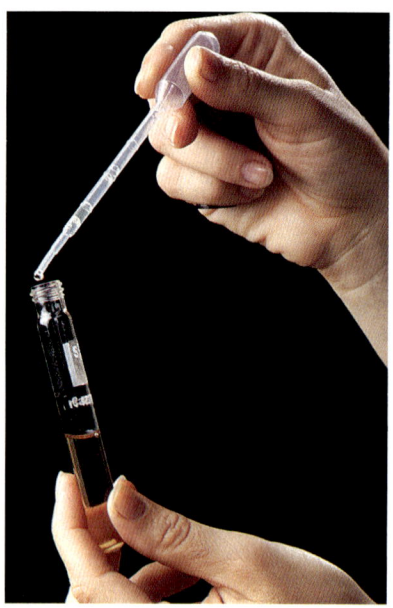

Figure 144.
Disks of the antibiotic to be
tested are added to a tube
containing broth medium.

Figure 145.
A few drops of the organism to
be tested are added.

The tubes are incubated anaerobically for 48 hr at 37°C. A tube with no antibiotic
disks added is inoculated to serve as a growth control, and a tube with only disks
added (no organisms) serves as a sterility control.

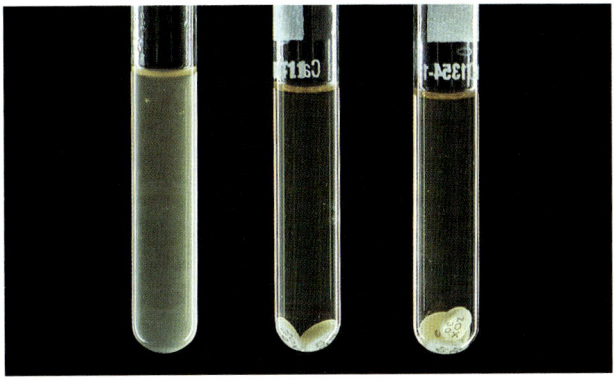

Growth control   Test organism   Sterility control

Figure 146.
If no growth is detected in the test, the organism is sensitive.

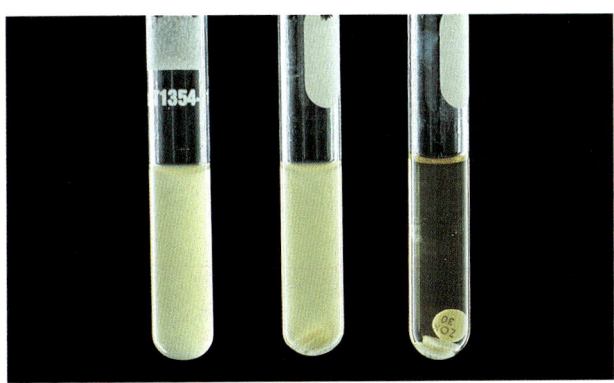

Growth control   Test organism   Sterility control

Figure 147.
If there is growth in the test, the organism is resistant.

Table 10. BROTH DISK ELUTION

Advantages	Disadvantages
Cost effective	Poor correlation with standard methods, particularly for the newer cephalosporins
Easy to perform	
Antibiotics need not be determined in advance	No MICs can be determined
Single isolates are easily tested	Background turbidity may make interpretation difficult

This method is no longer approved by the NCCLS.

Newer Methods

Newer, simpler techniques are being evaluated for anaerobic susceptibility testing:

The **spiral gradient endpoint system** deposits antibiotic in a continuous concentration gradient on agar plates. MICs may be determined by this technique. The advantages of the agar dilution method (good growth of fastidious anaerobes, convenient for testing many strains) are retained, and the system gains the time and cost advantages of simpler methods. However, the equipment required is expensive.

The **E-strip system** uses strips of antibiotics that form a continuous gradient when placed on an agar surface. After incubation the MICs may be read directly from the plates. There is not much experience with this method as yet.

Macrobroth techniques, which may provide for more versatile handling of fastidious organisms and more accurate MICs than microbroth techniques, are being developed.

Susceptibility Patterns of Anaerobic Bacteria

Until several years ago, anaerobic susceptibility patterns were relatively stable and predictable. Shifting patterns and variable efficacy of many of the newer antimicrobial agents have made consideration of the susceptibility patterns of anaerobes mandatory.

In vitro data may not always correlate with clinical response, as other factors (host defenses, surgical management, etc.) may be involved.

Bacteroides

Penicillin G, penicillin V, and ampicillin—effective against ∼ 5 to 20% of the *B. fragilis* group and ∼ 70% of other *Bacteroides* species

Other penicillins (nafcillin, isoxazolyl penicillins such as oxacillin, dicloxacillin)—active against much less than 70% of the *Bacteroides*

Aminoglycosides—generally ineffective against *Bacteroides* because these drugs cannot penetrate into the cell

Bacteroides fragilis Group

The combination **beta-lactam/beta-lactamase inhibitor** agents are effective against essentially all strains of the *B. fragilis* group

Imipenem, metronidazole, chloramphenicol—generally active against all strains of *B. fragilis* group species. (However, there have been reports of up to 5% imipenem resistance among *B. fragilis* isolates in Japan. Metronidazole-resistant strains of the *B. fragilis* group have also been reported. Metronidazole resistance is carried on a transferable plasmid and may spread to other strains.)

Third-generation cephalosporins are active against ∼ 50% for the *B. fragilis* group except for ceftizoxime, which may be active against 50 to 95% of this group, depending on the testing method

Bacteroides fragilis Species

First- and second-generation cephalosporins (except for cefoxitin and cefotetan)—active against ∼ 50% of *B. fragilis* strains

Cefoxitin and cefotetan—active against ∼ 90% of *B. fragilis*

Other *B. fragilis* Group Organisms

First- and second-generation cephalosporins (including cefotetan)—generally active against ∼ 50% of the group strains

Cefoxitin, clindamycin—active against ∼ 70 to 80% of the strains encountered in a number of institutions

Pigmented *Bacteroides* and *Porphyromonas*

Generally susceptible to most of the agents used in anaerobic infections

Bacteroides gracilis

One of the most resistant of the *Bacteroides* commonly involved in serious, deep-seated infections

First- and second-generation cephalosporins, penicillins, beta-lactam/beta-lactamase inhibitor combinations—active against ~ 60% of *B. gracilis*

Occasional strains of **imipenem** or **metronidazole-resistant** *B. gracilis* may be encountered

Other Non-*B. fragilis* Group *Bacteroides*

Generally susceptible to the drugs used to treat anaerobic infections

About 30% resistant to **penicillin G** and other **non-beta-lactamase-resistant beta-lactams**

Fusobacterium

Generally susceptible to **cephalosporins** (85 to 100%) and **penicillins**

Low level of resistance to **clindamycin**

Motile Anaerobic Gram-Negative Rods

Almost always susceptible to **chloramphenicol** and **metronidazole**

Variable in susceptibility to **beta-lactams**

Non-Spore-Forming Gram-Positive Rods

(*Eubacterium* sp., *Actinomyces* sp., *Propionibacterium sp.*, *Lactobacillus* sp.)

Susceptible to **beta-lactam/beta-lactamase inhibitor combinations**, most **cephalosporins,** and **penicillins**

Often resistant to **metronidazole**

Somewhat resistant to **moxalactam**

Clostridia

Clostridium difficile

Susceptible to **penicillins**

Resistant to **cephalosporins** (except **cefotetan**) in vitro, even when combined with beta-lactamase inhibitors

Often resistant to **clindamycin**

Caution—**ampicillin** and **cefotetan**, as well as **clindamycin**, may lead to overgrowth of *C. difficile* in the colon

Clostridium perfringens

Almost always susceptible to all of the agents commonly used for anaerobic infections

Other *Clostridium* species

15 to 30% resistant to **cefoxitin** and **clindamycin**, >30% resistant to **other cephalosporins**

Occasional resistance to **penicillin G** among several species

Chloramphenicol and **metronidazole** almost always active

Certain species (*C. ramosum*, *C. butyricum*, *C. clostridiiforme*) may produce beta-lactamases

Anaerobic Cocci (*Peptostreptococcus* sp.)

Generally susceptible to **beta-lactam/beta-lactamase inhibitor combinations, penicillins, cephalosporins**

Occasional **metronidazole-resistant** strains—usually microaerophilic

Table 11. SUSCEPTIBILITY OF ANAEROBES TO ANTIMICROBIAL AGENTS*

	B. fragilis Species	B. fragilis Group (excluding B. fragilis)	B. gracilis	Other Bacteroides	Fusobacterium
>95% susceptible	Ampicillin+Sulbactam; Piperacillin+Tazobactam; Ticarcillin+Clavulanate; Cefoperazone+Sulbactam; Imipenem; Chloramphenicol; Metronidazole	Ampicillin+Sulbactam; Piperacillin+Tazobactam; Ticarcillin+Clavulanate; Cefoperazone+Sulbactam; Imipenem; Chloramphenicol; Metronidazole	Imipenem; Chloramphenicol	Piperacillin; Ampicillin+Sulbactam; Ticarcillin+Clavulanate; Cefoperazone; Cefotaxime; Cefoxitin; Cefoperazone+Sulbactam; Ceftizoxime; Imipenem; Chloramphenicol; Clindamycin	Penicillin G; Piperacillin; Ampicillin+Sulbactam; Piperacillin+Tazobactam; Ceftizoxime; Imipenem; Chloramphenicol; Clindamycin; Metronidazole
85-95% susceptible	Cefotetan; Cefoxitin; Clindamycin; Piperacillin	Piperacillin	Metronidazole	Cefotetan; Ceftazidime; Ceftriaxone	Ticarcillin+Clavulanate; Cefoperazone; Cefotaxime; Cefotetan; Cefoxitin; Ceftriaxone; Cefoperazone+Sulbactam
70-84% susceptible	Ceftizoxime; Moxalactam	Cefoxitin; Clindamycin	Piperacillin; Ampicillin+Sulbactam; Moxalactam; Cefoperazone+Sulbactam	Penicillin G; Moxalactam	Ceftazidime; Moxalactam
50-69% susceptible	Cefoperazone; Cefotaxime; Ceftizoxime; Moxalactam	Cefoperazone; Cefotetan; Ceftizoxime; Moxalactam	Penicillin G; Ticarcillin+Clavulanate; Cefoperazone; Cefotaxime; Cefotetan; Cefoxitin; Ceftizoxime; Clindamycin		
<50% susceptible	Penicillin G	Penicillin G; Cefotaxime; Ceftazidime; Ceftriaxone	Ceftazidime		

PERCENT OF STRAINS SUSCEPTIBLE AT BREAKPOINT

>95% susceptible   85-95% susceptible   70-84% susceptible   50-69% susceptible   <50% susceptible

The ranking within a color group does not reflect degree of activity; drugs are arranged by classes.

*These data represent a compilation of studies from the Wadsworth Anaerobe Laboratory.

SUSCEPTIBILITY OF ANAEROBES TO ANTIMICROBIAL AGENTS*

Bilophila	Peptostreptococcus	Clostridium perfringens	Other Clostridium species	Nonspore-forming Gram-positive rods
Chloramphenicol Metronidazole	Penicillin Piperacillin Ampicillin+Sulbactam Ticarcillin+Clavulanate Cefoperazone Cefotetan Ceftazidime Ceftriaxone Moxalactam Cefoperazone+Sulbactam Imipenem Chloramphenicol	All drugs are active at ≥95% level	Amoxicillin Ampicillin Carbenicillin Penicillin G Piperacillin Ticarcillin Ampicillin+Sulbactam Imipenem Chloramphenicol Metronidazole	Penicillin G Piperacillin Ampicillin+Sulbactam Ticarcillin+Clavulanate Cefotaxime Ceftizoxime Imipenem Chloramphenicol
Clindamycin	Metronidazole			Cefotetan Cefoxitin Ceftazidime Ceftriaxone Cefoperazone+Sulbactam Clindamycin
Cefotaxime Cefotetan Imipenem	Clindamycin		Cefoxitin Moxalactam Clindamycin	Cefoperazone Moxalactam
Penicillin G Ticarcillin Ticarcillin+Clavulanate Cefoxitin Ceftizoxime			Cefoperazone Cefotaxime Ceftizoxime Ceftriaxone	Metronidazole
			Ceftazidime	

PERCENT OF STRAINS SUSCEPTIBLE AT BREAKPOINT

>95% susceptible	85-95% susceptible	70-84% susceptible	50-69% susceptible	<50% susceptible

The ranking within a color group does not reflect degree of activity; drugs are arranged by classes.

*These data represent a compilation of studies from the Wadsworth Anaerobe Laboratory.

Table 12. IN VITRO ANTIMICROBIAL SUSCEPTIBILITY OF MOTILE, ANAEROBIC GRAM-NEGATIVE BACILLI

Organism (approximate number of isolates)	Drugs Eliciting Indicated Result in Susceptibility Tests[a]		
	Susceptible	Variable	Resistant
Butyrivibrio (1)	Penicillin G, chloramphenicol, erythromycin, tetracyclines	—	Bacitracin, streptomycin, kanamycin, lincomycin, sulfonamides
Succinimonas (1)[b]	Bacitracin, oxytetracycline, penicillin G	—	Kanamycin, streptomycin, erythromycin
Succinivibrio (2)	Penicillin G, tetracycline, erythromycin, chloramphenicol	Clindamycin	—
Wolinella (19)	Metronidazole, clindamycin, chloramphenicol, tetracycline, erythromycin, imipenem, ciprofloxacin	Penicillin G, piperacillin, cephalosporins, gentamicin, rifampin, polymyxins	Vancomycin, bacitracin, nalidixic acid
Campylobacter (7)	Clindamycin, chloramphenicol, metronidazole, penicillin G, tetracycline, erythromycin	—	Vancomycin, bacitracin, rifampin
Desulfovibrio (1)	Penicillin G, clindamycin, chloramphenicol, tetracycline, erythromycin	—	Vancomycin, colistin
Selenomonas (32)	Clindamycin, chloramphenicol, metronidazole	Penicillin G, ampicillin, erythromycin, tetracycline	Vancomycin, colistin
Anaerobiospirillum (17)	Cephalothin, chloramphenicol, tetracycline, rifampin	Ampicillin, erythromycin, metronidazole, nalidixic acid	Vancomycin, trimethoprim, penicillin G

[a] Strains are reported as susceptible if > 90% of tested isolates were susceptible, as variable if 50 to 90% of tested isolates were susceptible, and as resistant if < 50% of tested isolates were susceptible. The susceptibility of all isolates was not reported for all antimicrobial agents. (From Johnson and Finegold, 1987.)

[b] Susceptibility data are for a strain isolated from an animal rumen.

Cautions in Interpreting Susceptibility Data: Key Questions to Ask

What Method Was Used?

Variations in testing methodology between laboratories may lead to confusion regarding resistance of anaerobes.

The MICs of some antimicrobials against anaerobes are highly method-dependent.

Microbroth and macrobroth dilution techniques usually give an MIC one twofold dilution lower than agar dilution techniques (and thus higher reported percent susceptible).

The NCCLS has recommended an agar dilution technique for use as a reference and standard method. Unfortunately, this method does not support the growth of many clinically important anaerobes. The Wadsworth Anaerobe Laboratory uses the NCCLS method, but with a blood-containing medium that supports the growth of most clinically relevant anaerobes.

Is There Clustering of MIC Values Near the Breakpoint?

Significant numbers (40 to 60%) of the strains tested have MICs within one dilution of the breakpoint for most beta-lactam agents (except ampicillin/sulbactam, imipenem), clindamycin, and chloramphenicol. These strains may be variably labeled as susceptible or resistant, WITHIN THE ALLOWABLE ERROR OF THE TECHNIQUE (one twofold dilution in either direction). This phenomenon may lead to significant variability in results. (See chart on page 142).

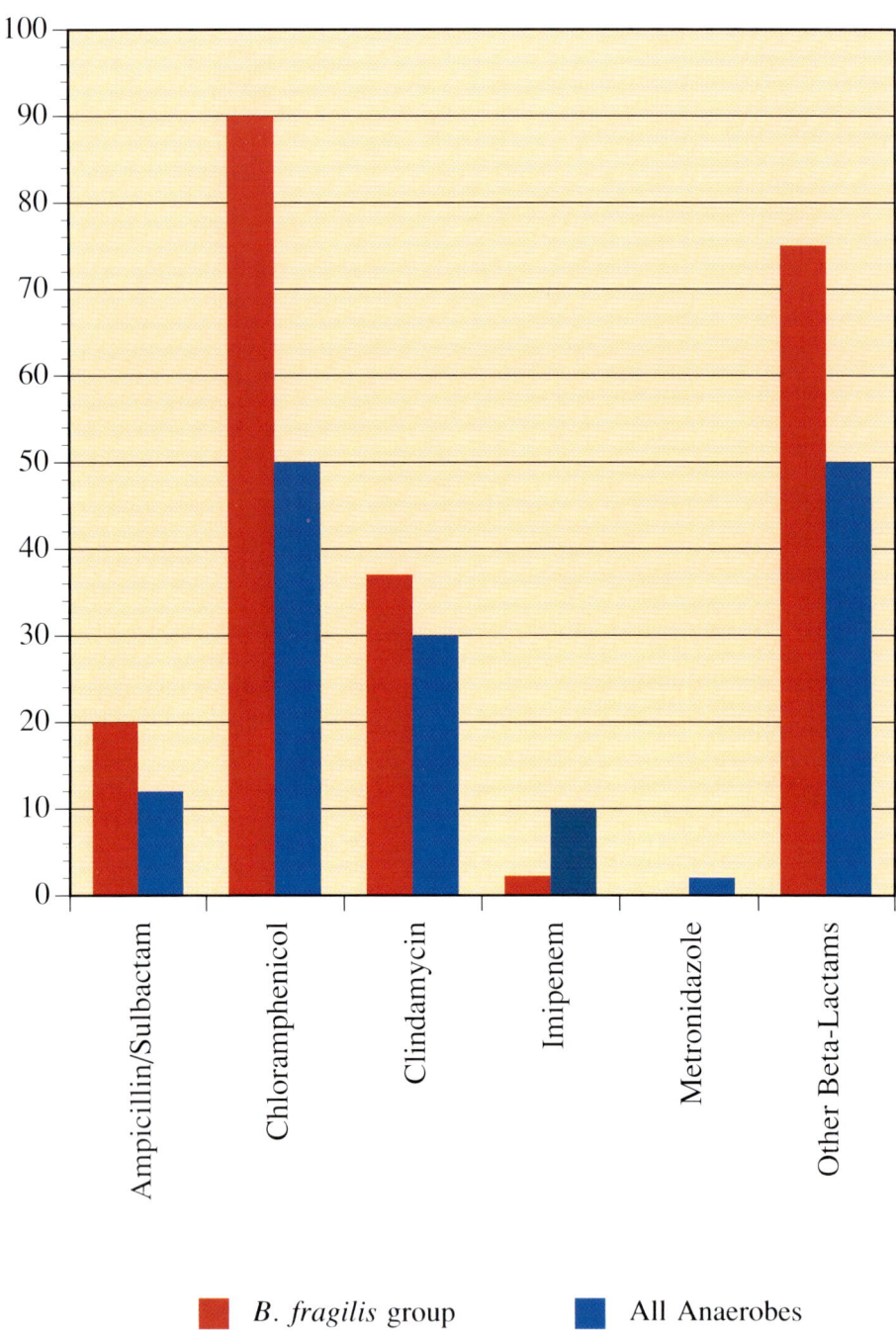

Clustering: % of Strains with MIC Values
within One Dilution of Breakpoint Values

How Is the MIC Defined?

The NCCLS currently defines the agar dilution MIC endpoint as no growth, one or two colonies, or a barely visible haze. There are no clear guidelines for reading or reporting results that do not fall neatly into these categories.

What Value Is Being Reported?

MIC 50s are often reported, and yet give little practical information about the usefulness of an antibiotic. MIC 90s are more predictive of the efficacy of an antibiotic, and percent susceptible at concentrations at or near the breakpoint probably provide the most information. When comparing studies, the same values must be considered.

What Breakpoints Are Being Used?

The NCCLS provides a list of approved breakpoints for determining susceptibility. However, these are not always used in the literature. Foreign regulatory committees often approve breakpoints that are different.

What Organisms Are Being Tested?

It is extremely important to differentiate between the species *B. fragilis* and the *B. fragilis* group. *B. fragilis* tends to be much more susceptible to both beta-lactam antibiotics and clindamycin than the other members of the group. Species other than *B. fragilis* account for approximately half of the strains isolated from infections involving the *B. fragilis* group so their susceptibility patterns must be taken into account. Furthermore, susceptibility patterns may change, so it is important to test relatively recent clinical isolates.

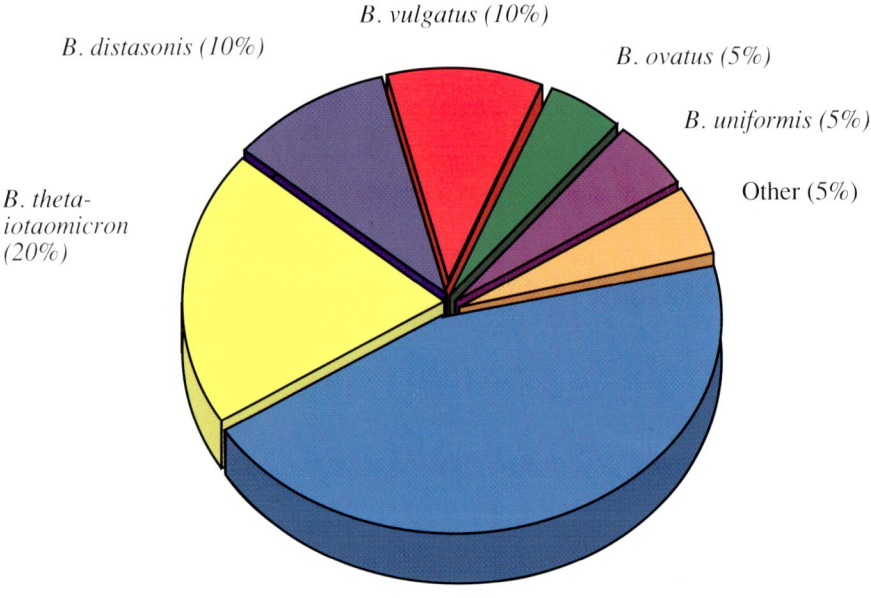

Mechanisms of Resistance in Anaerobes

Anaerobic bacteria develop resistance to antimicrobial agents just as nonanaerobes do, and the mechanisms of such resistance are very much the same. The problem with resistance among anaerobes is greatest with the gram-negative anaerobic bacilli. Inactivation (of beta-lactam drugs) by beta-lactamases is the single most common mechanism of resistance. This is fortunate because there are several beta-lactamase inhibitors that can block these enzymes and thereby restore effectiveness of the beta-lactam drugs against these organisms. Following is a listing of the major mechanisms of antimicrobial resistance in anaerobes.

Beta-Lactamase Production

Beta-lactamases have been found in most strains of the *B. fragilis* group, in some strains of other *Bacteroides*, in some strains of *Fusobacterium* and *Clostridium*, and may exist in other species as well.

There are also organisms that do not produce β-lactamases (e.g., some strains of *B. distasonis, B. gracilis,* and *Bilophila wadsworthia)*, yet are highly resistant to beta-lactam antibiotics.

Other Enzymes That Inactivate Drugs

Chloramphenicol may be rendered inactive by enzyme action.

Failure of Drugs to Penetrate Bacterial Cells

Aminoglyoside resistance is due to the absence of transport systems that can move the drugs into the cells.

Hydrophobicity of the drug (and hence its ability to permeate the cell membrane) has been correlated, in some cases, with its antimicrobial activity.

Drugs Pumped Out of Cells

This is the method of tetracycline resistance.

Changes in the Target of the Drug

Changes in penicillin-binding proteins may affect beta-lactam action.

Changes in the target ribosomal protein may be responsible for clindamycin resistance.

Differences in the dihydrofolate reductase target account for trimethoprim resistance in anaerobes.

Failure to Convert Drug to Active Form

Decreased activity of the enzyme(s) necessary to convert metronidazole to its active form is responsible for resistance to this agent.

Resistance Factors Transferred by Plasmids or by Conjugation

Bacteroides possess plasmids that code for both resistance and transfer of this resistance to clindamycin, erythromycin, and tetracycline.

Clostridium has plasmids accounting for transferable resistance to tetracycline and erythromycin.

Plasmids that can transfer metronidazole resistance have been demonstrated in *Bacteroides*.

Conjugation (direct transfer of genetic material from cell to cell) has been implicated in transfer of tetracycline and high-level ampicillin resistance.

Inducibility

Exposure to tetracycline has been shown to increase resistance to this agent in *Bacteroides*.

Selected References

1. Wexler H, Finegold SM. Antimicrobial resistance in *Bacteroides*. *J. Antimicrob. Chemother.* 19:143-146, 1987.

2. Finegold SM, Wexler HM. Therapeutic implications of bacteriologic findings in mixed aerobic-anaerobic infections. *Antimicrob. Agents Chemother.* 32:611-616, 1988.

3. Finegold SM, *et al.* Susceptibility testing of anaerobic bacteria: a minireview. *J. Clin. Microbiol.* 26:1253-1256, 1988.

4. Cuchural GJ Jr, Tally FP, Jacobus NV, et al. Susceptibility of the *Bacteroides fragilis* group in the United States: analysis by site of isolation. *Antimicrob. Agents Chemother.* 32:717-722, 1988.

SECTION 6

THERAPY OF
ANAEROBIC
INFECTIONS

SECTION 6

THERAPY OF
ANAEROBIC INFECTIONS

In most anaerobic infections (except pulmonary), surgical drainage and debridement are crucial for a favorable outcome. Other modes of therapy, such as hyperbaric oxygen and hydrogen peroxide, occasionally may be useful. Other than surgery, the major therapeutic approach is the use of antimicrobial agents. The only drugs at present that are active against essentially all anaerobes are chloramphenicol, imipenem, combinations of beta-lactam drugs and beta-lactamase inhibitors (ampicillin plus sulbactam and amoxicillin or ticarcillin plus clavulanic acid) and metronidazole. Metronidazole is not very active *vs.* gram-positive, nonspore-forming rods, such as *Actinomyces*. Drugs with lesser activity (e.g., clindamycin, cefoxitin, penicillin G) may be very useful in patients who are not seriously ill or in situations where susceptibility to these other agents can be demonstrated in vitro. Factors other than susceptibility of the microorganism must be taken into account when choosing an antimicrobial agent — toxicity, pharmacology (normal and in patients with impaired renal or hepatic function), ease of administration, central nervous system penetration, impact on the normal flora, bactericidal activity, and cost. See Tables 13–15: Comparison of Antimicrobial Drugs Active Against Anaerobic Bacteria, Toxicity and Side Effects of Drugs Most Active Against Anaerobic Bacteria, and Initial Dosage of Drugs Most Active Against Anaerobes.

Table 13. COMPARISON OF ANTIMICROBIAL DRUGS ACTIVE AGAINST ANAEROBIC BACTERIA

Drug	Aerobic Spectrum	Toxicity	CSF Penetration	Effect on Normal Flora	Dosage Form Available Oral	Parenteral	Bactericidal Activity
Penicillin	Poor	Low	Good	Minimal	Yes[a]	Yes	Very good
Clindamycin	Poor	Low-moderate	Poor	Major	Yes	Yes	Moderate
Metronidazole	None	Low	Excellent	Minimal	Yes	Yes	Excellent
Chloramphenicol	Good	High	Excellent	Minimal	Yes	Yes	None
Cefoxitin	Good	Low	Moderate	Moderate-Major	No	Yes	Very good
Carboxypenicillins[b]	Good	Relatively low	Good	Minimal	No[c]	Yes	Very good
Piperazine and ureidopenicillins[d]	Very good	Relatively low	Good	Minimal-moderate	No[c]	Yes	Very good
Imipenem	Excellent	Relatively low	Good?	Minimal	No	Yes	Very good
Ampicillin/sulbactam	Very good	Low	Good	Moderate	Invest.[e]	Yes	Very good
Ticarcillin/clavulanate	Excellent	Relatively low	Good?	Minimal-moderate?	No	Yes	Very good
Amoxicillin/clavulanate	Very good	Low	Good?	Minimal-moderate	Yes	No	Very good

[a] Parenteral form usually preferred for anaerobic infections: higher dosage and higher blood levels are facilitated.
[b] Carboxypenicillins—carbenicillin, ticarcillin.
[c] Poorly absorbed by the oral route and therefore not suitable for therapy of systemic infection.
[d] Piperazine penicillin—piperacillin. Ureidopenicillins—mezlocillin, azlocillin.
[e] Investigational

Table 14. TOXICITY AND SIDE EFFECTS OF DRUGS MOST ACTIVE AGAINST ANAEROBIC BACTERIA[a]

Drug	Hypersensitivity Reactions	Neurotoxicity[b]	Bleeding[c]	Pseudomembranous Colitis	Aplastic Anemia	Fluid and Electrolyte Problems
Penicillin	●	●				
Ampicillin	●	●		●		
Clindamycin				●		
Metronidazole		●				
Chloramphenicol					●	
Cefoxitin	●			●		
Carboxypenicillins	●	●	●			●
Piperazine and ureidopenicillins	●	●	●			
Imipenem	●	●				
Ampicillin/sulbactam	●					
Ticarcillin/clavulanate	●	●	●			
Amoxicillin/clavulanate	●					

[a]Only more commonly encountered reactions are noted. For example, pseudomembranous colitis may be seen with all of the agents listed, but the incidence is distinctly higher with the three agents marked.

[b]Seen with high and/or prolonged dosage, usually in patients with impaired ability to excrete and/or conjugate the drug. Patients with a history of seizures may be predisposed.

[c]Due to either prolonged prothrombin time (vitamin K deficiency) or impaired platelet function.

Table 15. INITIAL DOSAGE OF DRUGS MOST ACTIVE AGAINST ANAEROBES[a]

Drug	Route of Administration	Dosage
Penicillin G[b]	Intravenous	10-20 million units/day
Piperacillin[c]	Intravenous	3-4 g every 4-6 hours
Cefoxitin	Intravenous	1-2 g every 4-6 hours
Chloramphenicol[b]	Intravenous[d] or oral	1 g every 6 hours
Metronidazole[b]	Intravenous or oral	1 g loading dose (IV only), then 500 mg every 6 hours
Clindamycin[b]	Intravenous	600 mg every 6 hr or 900 mg every 8 hours
	Oral	150-300 mg (up to 450 mg) every 6 hours
Imipenem	Intravenous	0.5-1.0 g every 6-8 hours
Ampicillin/sulbactam	Intravenous	1.5-3.0 g every 6 hours
Ticarcillin/clavulanate	Intravenous	3.1 g every 4-6 hours
Amoxicillin/clavulanate	Oral	250-500 mg every 8 hours

[a] Assuming an average-size adult (with normal renal function in the case of penicillin, piperacillin, and cefoxitin).
[b] Dosage may be reduced later in the treatment course in many patients.
[c] Or other ureido- and carboxypenicillins.
[d] Administer each dose over 30 minutes.

Summary of Usual Bacteriology and Examples of Appropriate Therapy* for Three Major Types of Mixed Anaerobic Infections

I. Pleuropulmonary Infections (aspiration pneumonia, lung abscess, empyema, etc.)

    A.   Community-acquired disease (Indigenous flora)

        1.   Pathogens

            a)   Anaerobes

                (1)   *Bacteroides*

                    (a)   Pigmented *Bacteroides* (and *Porphyromonas*)

                    (b)   *B. oris*

                    (c)   *B. buccae*

                    (d)   *B. oralis* group

                    (e)   *B. ureolyticus* group (especially *B. gracilis*)

                    (f)   *B. fragilis* group

                (2)   *Fusobacterium*

                    (a)   *F. nucleatum*

                    (b)   Others

                (3)   *Peptostreptococcus* sp.

            b)   Viridans streptococci

        2.   Therapy

            a)   Mild to moderately severe infection: penicillin G 10-15 million U/day is first choice. Failures may be treated with the addition of metronidazole or clindamycin or with ampicillin/sulbactam.

            b)   Severe infection or host with significant other illness: ampicillin/sulbactam or penicillin plus metronidazole.

*Not every possible regimen is listed. Choice will depend on seriousness of infection, general health of patient, specific organisms recovered and their resistance pattern, etc.

B. Hospital-acquired disease (Nosocomial pathogens); above indigenous flora is also involved in hospital-acquired disease in addition to organisms listed below.

    1. Pathogens

        a) *Staphylococcus aureus*

        b) Gram-negative bacilli

            (1) *E. coli*

            (2) *Klebsiella*

            (3) *Enterobacter*

            (4) *Serratia*

            (5) Other *Enterobacteriaceae*

            (6) *Pseudomonas*

    2. Therapy

        a) Mild to moderately severe infection: ampicillin/sulbactam or cefoxitin

        b) Severe infection, host with significant other illness, or treatment failure: ticarcillin/clavulanate or imipenem ± amikacin and/or vancomycin (depending on flora)

II. Intra-abdominal Infections

    A. Appendicitis, diverticulitis, infection following bowel surgery, etc.

        1. Pathogens

            a) Aerobic, facultative organisms

                (1) *E. coli*

                (2) *Enterococcus*

                 (3) Viridans streptococci

                (4) *Staphylococcus aureus**

                (5) *Klebsiella, Enterobacter**

                (6) *Pseudomonas**

*Seen particularly in hospitalized patients receiving antimicrobial agents

b)   Anaerobes (outnumber aerobes, facultatives by approximately three to one)

      (1)   *Bacteroides fragilis* group (especially *B. fragilis, B. thetaiotaomicron*)

      (2)   Other *Bacteroides*

      (3)   *Bilophila wadsworthia*

      (4)   *Peptostreptococcus*

      (5)   *Clostridium perfringens*

      (6)   Other clostridia

2.   Therapy

    a)   Appendicitis with only local complications at most: ceftizoxime, cefoxitin, or cefotetan

    b)   More significant infections: clindamycin plus aminoglycoside ± penicillin or ampicillin

    c)   Serious intra-abdominal infection (other than biliary tract infection), host with significant other illness, or treatment failure: ticarcillin/clavulanate, imipenem, or ceftazidime plus metronidazole (± oxacillin or vancomycin), all ± amikacin

B.   Biliary tract infection

1.   Pathogens

    a)   Uncomplicated

      (1)   *E. coli*

      (2)   *Klebsiella*

      (3)   *Enterococcus*

      (4)   *C. perfringens* (5% of patients)

    b)   Complicated

      (1)   *B. fragilis* group also involved

2.   Therapy

    a)   Uncomplicated: ampicillin or ampicillin/sulbactam

    b)   Complicated: ampicillin/sulbactam

III. Female Genital Tract Infection (pelvic inflammatory disease, endometritis, etc.)

    A. Pathogens

        1. Aerobes, facultatives

            a) Streptococci

                (1) Groups A, B

                (2) *Enterococcus*

                (3) Viridans streptococci

            b) Gram-negative rods

                (1) *E. coli*

                (2) *Klebsiella*

            c) Gonococcus

            d) *Chlamydia*

            e) *Mycoplasma hominis*

        2. Anaerobes

            a) *Peptostreptococcus*

            b) *B. fragilis* group

            c) Other *Bacteroides*

                (1) *B. bivius*

                (2) *B. disiens*

                (3) Pigmented *Bacteroides* (and *Porphyromonas*)

                (4) Others

            d) Clostridia

                (1) *C. perfringens*

                (2) Others

            e) *Actinomyces, Eubacterium, Propionibacterium propionicus*

B. Therapy

   1. Mild to moderately severe infection: ceftizoxime, ampicillin/sulbactam, cefoxitin, clindamycin plus gentamicin, or cefotetan (all ± tetracycline or other agent for STD)

   2. Severe infection, host with significant other illness, or treatment failure: ampicillin/sulbactam, ticarcillin/clavulanic acid, imipenem, ampicillin plus metronidazole